CLIMATE CHANGE, INTERRUPTED

CLIMATE CHANGE, INTERRUPTED

Representation and the Remaking of Time

Barbara Leckie

Stanford University Press
Stanford, California

STANFORD UNIVERSITY PRESS

Stanford, California

©2022 by Barbara Leckie. All rights reserved.

No part of this book may be reproduced or transmitted in any form or by any means, electronic or mechanical, including photocopying and recording, or in any information storage or retrieval system without the prior written permission of Stanford University Press.

Printed in the United States of America on acid-free, archival-quality paper

Library of Congress Cataloging-in-Publication Data

Names: Leckie, Barbara, author.

Title: Climate change, interrupted : representation and the remaking of time / Barbara Leckie.

Description: Stanford, California : Stanford University Press, 2022. | Includes bibliographical references and index.

Identifiers: LCCN 2022004390 (print) | LCCN 2022004391 (ebook) | ISBN 9781503633070 (cloth) | ISBN 9781503633988 (paperback) | ISBN 9781503633995 (ebook)

Subjects: LCSH: Climatic changes in literature. | Time in literature. | Social action in literature.

Classification: LCC PN56.C612 L43 2022 (print) | LCC PN56.C612 (ebook) | DDC 809/.9336—dc23/eng/20220712

LC record available at https://lccn.loc.gov/2022004390

LC ebook record available at https://lccn.loc.gov/2022004391

Cover art: Still from *Incoming*, 2017, immersive video installation, 52 mins 12 seconds. © Richard Mosse. Courtesy of the artist and Jack Shainman Gallery

Cover design: Rob Ehle

Typeset by Newgen in Adobe Garamond Pro 11/14.4

for Michal and Ben, again and always

Table of Contents

Preface: A Gaping Hole in the Line   ix

1 **About Time: Three Beginnings**   1
   Interruption   3
   Post-time   27
   Alarming!   54

2 **On Time: Four Experiments**   79
   Layering   81
   In the Idiom of the Self-Help Guide   126
   Found Questions   145
   FrankenClimate   156

3 **Meanwhile: Two Endings**   181
   The Academic Book, Interrupted   183
   Unfinished   190

Acknowledgments   191
Notes   197
Bibliography   223
Index   241

Preface

A Gaping Hole in the Line

In April 2019 Ann Cognito and her service dog began a two-thousand-mile trek from Calgary to Ottawa. It took them seven months. Cognito made this arduous journey across Canada to protest the government's lack of action on the climate crisis. When she arrived in Ottawa, where I live, she pitched her tent on what is little more than a traffic island across the street from Parliament. She soon gained the support of local environmental and climate groups, and other tents joined hers on the small snow-covered junction. Cognito planned to continue camping there until the prime minister listened to, and acted on, her demands. When I visited her on a bright winter day in December, banners that read "Climate Emergency Camp" and "Walk to Waken the Nation" hung from the orange, blue, and red nylon tents that were zipped tightly against the cold. It was -30°C, and I was there to bring some supplies and a thermos of tea.

Voices came from one of the tents. I called out, Hello? Someone answered and, beginning to unzip the front flap, invited me to join them. I declined. It was cold, I had work to do, and I didn't have time to sit and talk. I wanted to participate but I didn't want to enter their small tent. I didn't say that though; I said that I didn't want to let too much cold air in the tent and I'd just pass them what I'd brought. A hand reached out and took the things, and the voices from within thanked me. We talked briefly.

They said they were cold despite the fact that they were huddled closely together. I inwardly congratulated myself on my decision not to enter. I'm an academic; I sit alone at my desk and write and read. When I do activist work, I don't tend to huddle in small spaces with strangers. But as I've been writing this book I've often thought of Ann Cognito's long walk, the climate emergency camp she set up in Ottawa, and that winter day when I visited.

What would it have meant to be in that tent? Cognito herself was there to participate in the much larger figurative tent of the Canadian government. In general, it's better to be "under the tent" than outside, to be part of the decision-making group than watching from the wings. In an interview Cognito said that there was not much she could do in response to the climate crisis, but she could do *this*, walk across the country, set up camp. I began this book about a year before Cognito began her walk. In my case, writing is slower than walking, its slowness jars with and dismantles my focus on *emergency*. But like Cognito, I think to myself, I can do *this*. But what is *this*? And how does it add up to *that*, the systemic change necessary to confront the climate crisis?

Climate change, of course, is an established scientific fact. The climate change *idea* is the way in which climate change is understood by others, discussed, debated, and addressed.[1] The climate change idea is distinguished from other new ideas in at least three ways: it poses formidable challenges to representation; it demands a rethinking of the temporal modes on which the Global North has relied since the nineteenth century;[2] and it presents an existential risk to human and other species. The existential risk to living beings motivates a desire to confront the crisis, but the joint challenges posed by representation and time impede the clarity of that confrontation. Another way to put this point is that the climate change *idea* lacks a robust conceptual apparatus to address climate change; in other words, it is still in the process of being shaped. Rob Nixon names *slowness* as one aspect of climate change that contributes to its representational intractability.[3] The incremental, attritional character of the climate crisis, with environmental and earth changes taking place across decades and centuries, makes it difficult to represent climate change with the clarity of more immediate and sensational crises.

In this book, I build on Nixon's insight and extend it to the temporal modalities that have dominated in the Global North—the area of the globe most responsible for fossil fuel emissions—since the rise of the Industrial Revolution.[4] I adapt an idea of time from Walter Benjamin to focus, like Nixon, on one aspect of temporality—in my case, interruption. I ask what a supple and capacious understanding of interruption might offer a crisis stymied by "slow time" and chronological timelines invested in progress.[5] While I will develop the idea of interruption in more detail in what follows, Cognito's climate emergency camp offers me one example of what interruption can look like. In short, I should have gone inside the tent.

When I think about standing outside the climate emergency camp that cold day in December, I know that part of my struggle to understand the climate crisis is to find the words—or mode—to capture its staggering dimensions. I can't view this crisis from a distance. I am *in it*. Of course, in the last half century or so, academics have recognized the role that preexisting contexts play in shaping perception. I do not mean that I am *in it* in that way (although I am). Rather, what I am referencing here is being *in it*, in *something*, for which I do not yet have words. I could ignore that. I could carry on using the words and categories I do have and that have served me relatively well thus far. But I want to interrupt myself, I want to interrupt the patterns to which I turn, the conceptual models on which I rely, the tools that give me comfort and reassure me that I'm participating in a larger conversation. I'm enough of a product of my times and training to believe that there is no stepping outside those patterns and models. But I can pause. I can stop and listen. I can think about what other possibilities reside in the works I consult, the words I use, the work I do. In this spirit, I want to write with and through time, inhabit the nineteenth century in which my area of scholarly expertise lies together with the twenty-first century in which I live. I want to ask whether rethinking time can help me to address climate change more adequately and in the process contribute both to an idea in the making and a making of it otherwise.

These reflections take me to a stormy afternoon in late December 1879, 140 years before Cognito set up her tent, when passengers in Edinburgh waited to board the last train to Dundee. Unlike Cognito, these passengers were availing themselves of the newest technology. They were about to make a journey across the longest bridge in Europe, completed only the previous year. The Tay Bridge was a stunning feat of engineering éclat and the subject of many paintings, drawings, and letters home. The Tay River itself was a busy transportation channel, and before the bridge was built, a ferry shared the two-mile-wide river with steamboats, fishing vessels, sailboats, and sightseeing rowboats in addition to, for several months a year, the whaling ships that brought their bounty to the busy port to be processed and distributed.

At 7:14 that December evening, the signal box from the other side of the Tay River announced the train's approach. But the train did not arrive. Instead, as Walter Benjamin puts it in his 1932 radio broadcast for children, "the train simply plunged into the void" (*Selected Writings*, 2:566). Eyewitness accounts describe a gale strong enough to break windows and dislodge chimney pots. They describe an intermittently moonlit night, winds that drowned out all other sounds as families stepped outside to watch anxiously for the train due to arrive in Dundee. Several people also describe seeing lights that blinked and then suddenly vanished.[6] Seventy-seven lives were lost in the railway accident that night. When another train was sent to investigate the accident, Benjamin writes, it cautiously proceeded along the bridge tracks "until the driver applied the brakes so fiercely that the train almost leaped from the track. The moonlight had enabled him to see a gaping hole in the line" (567).

Climate Change, Interrupted is about a gaping hole in the line.[7] I take the Tay Bridge disaster as a point of departure for thinking about both lines and their interruption. What might it mean to hold the line *and* to interrupt it? Can the broken line be confronted not as a problem to be fixed but as an invitation to thinking otherwise? What might it mean to *stay with*, in Donna Haraway's sense, the gaping hole? In my comments here, the gaping hole in the line refers to a hole in the railway line and, as such, invokes industrial modernity writ large and, more recently, the Anthropocene. My own reading embraces these approaches and expands them to address both linear time and the written line. I'm interested in thinking about the many different temporal modes that linear time has displaced—indeed, sidelined—since industrial modernity. How do linear time and the written line work in tandem, and how might they work differently to experiment with and envision new possibilities? One decidedly nonexperimental response to the collapsed bridge was William McGonagall's poem, "The Tay Bridge Disaster." But the accident also drew the attention of Theodore Fontane. Benjamin cites a section from his poem, "The Bridge by the Tay," as follows:

> And the watchman's people full of alarm
> Anxiously gaze out through the storm.
> For the wind in furious play seemed to grow
> and down from the sky flames did glow,
> Bursting in glory as they descended
> to the water beneath . . . and all was ended. (*Selected Writings*, 2:566)

Like the "watchman's people," I watch, full of alarm, as the storm gathers strength. But then that last sentence arrives with its breathtaking ellipses. The ". . ." is the gaping hole in the line. It is the pause before "all was ended." Often, in climate change analyses, we rush to the dark end, read our own apocalyptic scenarios as harbingers of the end to come, and seek to ward off that full stop.[8] Indeed, the climate crisis has generated a huge store of books about endings—living with the end, end times, apocalypse and so on.[9] This book does not disavow or deny that catastrophe. That catastrophe is experienced today by the many people in refugee camps; in places where droughts persist for years; in places where flooding has taken away homes and livelihoods; in places where fires destroy and blacken beloved landscapes, homes, animals, and people. But it seeks to orient it, or tune it, to a different modulation of time that does not invest so heavily in "the end." Indeed, that sees endings otherwise: as plural, multiple, and always unfinished—the ending as provisional, a pause that stays with the ". . .". The ellipsis registers an interstitial, still-moving space, and that space invites us to think, too, about interstitial time, not only a line charging toward its end, but also the many possibilities for inhabiting time that the line on its own obscures.

In the days, weeks, and years that followed the Tay Bridge disaster, it was the subject of newspaper articles, poems, inquiries, and stories as people sought, through words and images, to understand it, to frame it in a way that offered some meaning. These accounts were part of a collective effort to find intelligibility for an event that had, that night, fractured understanding so grievously that it continued to ripple out across the decades. Indeed, as I have noted, in 1932, almost fifty years later, Benjamin was still returning to it in his radio address. And I am returning to it now in another moment of uncertainty as we again find before us a gaping hole in the line.

Each of these repetitions redoubles the line and, in ways I want to explore in this book, undoes it. I invoke the line to name the temporal mode—linearity and its accompanying commitment to progress—that has, in many ways, organized temporal thought in the Global North over the last two centuries.[10] But I don't want to lose sight of the fact that the line is what is in question here and the hole is a way of naming that space before questions are answered, concepts begin to form, and new ideas come into view. Although here, too, language defies me. For the gaping hole is also not *before* these shifts at all but always there, a disturbance, a disquiet, an unsettling. The capacity to define terms and articulate strategies affords individuals, communities, and nations a great deal of power. But the capacity to stay with what resists naming is its own power.[11] The capacity to recognize the provisional and improvisational aspect of all naming is perhaps where another power resides, a power that allows and respects the gaping hole, a pause, possibility, stutter, and space for imagining otherwise.

My argument in this book is that climate disruption is a gaping hole in the line now. Like those witnesses to the Tay Bridge disaster, we also seek stories that can make sense of where we are, that can close the hole, and allow us to build better, more secure bridges, cross wider rivers, and find safer modes of doing so. But unlike the Tay Bridge disaster, our gaping hole in the line is not confined to a moment in time, is not something for which solutions are readily at hand, is not, indeed, something that can even be adequately named. Of course we have names like *climate change*, *climate crisis*, and *climate disruption*. But for most people in the Global North, these names are only partially adequate to the conditions—the gaping hole—that currently confront us.[12] As in the Tay Bridge disaster, commentators try to find ways to define our crisis that will enable at once a processing of it and a way to address it. These responses inevitably draw a frame around where we are that allows it to feel more intelligible and manageable, that gives it a shape and a substance. This book, by contrast, pauses with the ellipses, stays with the gaping hole in the line—what I will variously call *interruption* and *post-time* in the pages that follow—to work the gaping hole and the alternatives to linearity that it suggests. The first half of this book does so with three essays that remain within the parameters of the academic albeit with a few nods to a different accent or note. The second half offers a suite of four experiments offering approaches that are more fully experimental.

Ann Cognito's climate emergency camp also resonates with the much larger Occupy movement begun in 2010. People camped out in New York's Zuccotti Park to protest the stark economic inequalities that had not been prevented—indeed, were abetted—by democratic principles and practices. The movement, tellingly, gained its name from *Adbusters*, a magazine that focused on capitalism's erosion of democratic ideals, and the replacement, in Ailton Krenak's words, of the citizen with the consumer (30). In particular, the slow fraying of ideas of democratic equality were starkly captured in the divide between the "1 percent" and the "99 percent"—terms the Occupy movement succeeding in bringing indelibly into public debate.

The Occupy movement's genius was to recognize the radical loss of urban public space and how camping in newly privatized and corporatized spaces made this loss instantly legible. The movement *used* space to make a larger point, the point of the increasing gulf between the rich and the poor that is now remembered as the leitmotif of the Occupy movement. That is, the movement took something that had happened to public space—its privatization and its incorporation—and refused it. It did so in the most basic and provisional way possible: by camping in that space. There is much to be said about tents being erected in the city and the way that this tactic offers a commentary at once on homelessness, the rising refugee encampments emerging around the world, and the conceptual divide between city and country. But I want to move in a different direction.

I want to think of provisional urban encampments in terms of time. If capitalism erodes free public space, it also erodes "free" time.[13] What might it mean to "occupy" time? When people camped in capitalist spaces, they sought to revise capitalist tenets through their actions. They changed how these spaces were understood by living in them. Within the existing world, they built something new. To be sure, the Occupy movement's contribution and the conversation it generated seems almost comically ineffectual now as we consider, over a decade later, an even greater economic gap and an even more attenuated democratic structure. But it is for these very reasons that a new orientation to time seems critical.

To occupy time differently could mean this: living with, staying with, the gaping hole instead of suturing it over and returning to the line; living with, staying with, the interruption.[14] Benjamin develops his theory of interruption, in part, to contest and revise histories of progress. He wants to write history differently, alert and attuned to the voices that have been excluded from mainstream histories. But, importantly, his project is not one of historical recovery, of looking back in time, finding those figures and restoring them to the narrative. Instead, his project is one that seeks to collapse, superimpose, or constellate time, to make time and revolution register, as he puts it, "quite otherwise" (*Selected Writings*, 4:402). This interruption occupies time differently, in ways that defy representation but that nevertheless demand acknowledgment. In this context, the gaping hole generates a reconstellation of time in which new possibilities, a living otherwise, emerge. The climate crisis is our gaping hole in the line today, a crisis so profound and deeply seated that it dislodges and upsets long-established responses to crises. It does so, in part, because the climate crisis itself involves a refiguration of temporality, a triggering of the interruption that helps us to think otherwise. That said, it is not usually addressed as such. But to evade the gaping hole is not only to sidestep or evade a vital component of what we confront as a species today but also to miss its invitation to inhabit time differently and, with it, to imagine responses to the crisis calibrated to a different experience of time.

Ann Cognito's emergency climate camp came to a sudden close. After enduring a hard, cold winter with night temperatures often plunging below -20°C, after talking to reporters and fellow activists but not to the prime minister, Cognito folded up her tent in late March 2020. The Covid-19 pandemic eclipsed the climate emergency. And yet they are both different faces of the same emergency.[15] Like climate change, the pandemic was a global event that threatened all people, but its impact was more grievous and brutal for the poor and marginalized. And like climate change, the pandemic reminded humans of their vital connections with each other and with other species.[16] Climate change and the pandemic are also entwined in ways that extend beyond these formal parallels. As scientists have illustrated, the pandemic is most likely the result of the climate crisis.[17] It has, pace the focus of this book, interrupted daily life around the world and, in many cases, reoriented the ways things are seen and done. It has also offered an unexpected and unparalleled example of global partnerships forged in the face of a shared crisis. To be sure, tempers have also flared, patience has frayed, and critiques of the global response have been issued from all corners. But the world has also received a glimpse of what a united response might look like as the climate crisis continues to deepen.

The pandemic forced Cognito to pack up her camp. For me, it had a different impact. I was writing about interruption—its importance and its possibility—when the world as many people knew it was interrupted. In this book I don't seek repairs or retreats, better or stronger bridges, figurative or otherwise, or carrying on along the same lines, whether those lines are iron rail lines of trains, the flight lines of planes, or written lines. I seek, instead, something that is more like a makeshift tent: provisional, conversational, unexpected, unfolded, and open to all. I want to think about different ways of working the gaping hole in the line, of mobilizing interruption, of living time. I also want to consider more closely the climate change idea that is emerging now through stories, debates, discussions, images, as well as silences and what we struggle to represent. It is not only that climate change is difficult to represent because of its slow violence; it is also difficult to represent because it animates a gaping hole in the line. It reminds us, indeed, that we are all the figurative "watchman's people," in the midst of the storm, watching as it gathers strength.

. . .

CLIMATE CHANGE, INTERRUPTED

PART I

About Time: Three Beginnings

About Time: First Beginning

Interruption

This book addresses the role of interruption, as a theory and a practice, in response to the climate crisis. To do so, it rethinks the temporal models indebted to origins, linearity, progress, and teleology on which the Global North typically relies to address climate change questions. These temporal models assign origins to climate change as well as devastating ends.[1] For many, the grim prognosis *is* the point around which to galvanize action. The series of intergovernmental reports on climate change culminated, most recently, in the 2021 Intergovernmental Panel on Climate Change Report's conclusion that "unless there are immediate, rapid and large-scale reductions in greenhouse gas emissions, limiting warming to close to 1.5°C or even 2°C will be beyond reach." The gravity of the climate crisis inescapable. The desire to avoid the destruction of the planetary habitat that supports the life of humans and other species, in short, should prompt us to act. The problem is that it doesn't. The Global North needs compelling alternatives to the way in which the climate change story has been told, alternatives that rethink the chronological and representational models on which it has relied. This book explores interruption itself as one such alternative approach.

. . .

And yet: I am writing at a moment when many are questioning the academic format of the book and what it delivers. Concerned about climate change and the timeline it asks us to consider, what impact can an academic book have? It is not only that academic books take a long time to write but also that they take a long time to publish. If I seriously feel that climate change needs to be addressed, then writing a book that will, at best, not be read by the public for three years is surely not the most expedient route to take. Worse, it implies that I don't fully endorse the idea of a climate crisis; my practice does not line up with my theory. Isn't my time better put elsewhere? Further, the book itself has a carbon price tag.[2] On what grounds, then, does it make sense to write an academic book on climate change?[3] There are many ways to respond to this question. For me, three rationales underpinned my writing even if I sometimes also questioned them: my commitment to thinking as action; my interest in exploring new forms for the academic book, especially in this climate-imperiled moment; and my sense, however inchoate, that an approach to climate change that pivoted on interruption might offer something not found elsewhere.

As a scholar located in the humanities, I am inclined toward what I will call here modalities of thinking.[4] I value the role of *thinking* to address seemingly intractable impasses like the current climate crisis. In these uncertain times, it seems important to develop modes of thinking that depart from, or at least do not cleave so closely to, traditional formats. It seems important also to reconnect thinking to the conditions—historical, political, material—that make it possible. Thinking requires time and it is also alert to, and inseparable from, form, mediation, and materiality. Consider, for example, the imprint of *race* in Lauret Savoy's *Trace*, a book that traces the elided history of race in environmental writing. Consider the difference between a telegram, a letter, a postcard, and an email, even when the words they communicate are identical. In the context of climate change, like many other critics, I'm interested in recalibrating the sciences, social sciences, and humanities in an effort (1) to generate a more serious, supple, and sophisticated dialogue between the different disciplines; (2) to acknowledge the ways in which each of the disciplines is *already* bound up with its others and to bring this entwinement into conversation; and (3) to showcase what the humanities, in particular, have to offer our current moment. If one of the obstacles to a successful climate response in recent years has been an ongoing division between the sciences and the humanities, the humanities confront that division and reject it (this is point 2 above).[5] Or they can. Humanities disciplines are nicely equipped to bind together disparate areas, to remind us they have always been connected, and to demonstrate the ways in which divisions are produced that sometimes serve a useful purpose and sometimes do not.[6] And the humanities are also where the gaping hole finds its most hospitable home.

The humanities, in other words, are good at thinking about thinking. If this sounds like one more way to stall climate change action, getting mired in abstraction and distancing one's self from doing anything, it is not. For the humanities have also demonstrated that thinking itself is a form of doing. In a recent book, Astra Taylor writes that democracy is a theory and a practice (88).[7] And, importantly for me here, the practice is part of the theory. Democratic theory is realized in practices that will always improvise and adapt to situations and feed back into the theory to create something new. So, too, thinking. I want to exercise a *practice* of thinking, which is also a practice of conversation and co-writing. We can never know what the humanities can contribute to the climate change impasse without trying out a range of possibilities, a practice in which, happily, many scholars are now engaged. There is a value in participating in those conversations and practices even when, especially when, we're not sure where our thinking will lead.

> ***Waypoints:*** *I've come to the field of climate change studies relatively late in my career. Often that sense of lateness is beset by the typical academic concerns—I don't know as much as others, how could I possibly "catch up," and so on—but sometimes that lateness also has a different register. This morning, for example, before sitting down to write this chapter, I heard on the radio that a space the size of 2,000 football fields is now being burnt in the Amazon rain forest each day. The rain forest not only absorbs carbon dioxide but also provides 30 percent of the world's oxygen. Jair Bolsonaro, however, was elected president of Brazil in January 2019 because he promised to develop the land, and farmers interviewed on the radio said, not unreasonably, that their livelihoods depended on clearing the trees so that they could farm. It reminded me of the debate so frequently heard in Canada between those who want to stop oil pipelines and those who depend upon their expansion for jobs to feed themselves and their families. This account of the rain forest burning—the reporter described thick plumes of smoke filling the sky—is only one of the countless contests between planetary*

> *survival and jobs that are occurring almost everywhere. When I hear it (and before I begin to interpret it), I am stunned. I feel ever more acutely that I am, we are, too late to respond adequately to climate change; that we have not established the practices of working together, collaboratively, to do whatever is necessary to offset further harm to existing ecosystems; that we are carrying on in a surreal space (as I am here) as if we can still do all the things we have usually done, write books, teach classes, attend conferences. That capitalism, capped or not, will swallow it all and grow bigger. This is a detour. My belatedness punctures the present moment in which I write. Meanwhile, fields and fields of old-growth rain forest continue to burn.*

If my first rationale for this book is my investment in thinking, my second is an investment in thinking that is extended and amplified through an exploration of different, less academic forms. The humanities and their habits of broad-ranging thinking are nicely equipped to navigate between the different approaches now required—political, economic, social, artistic, energy-related, scientific, and so on—and to remind us of the stakes at play in how we formulate the problem. So much of that work has been beautifully done. This book could not have been written without traditional academic books, but I am interested in something else.[8] Scholars often undervalue what other approaches can accomplish, and even what they constitute. And, for the most part, academics haven't begun to explore the many possibilities for alternative imagining and writing. I want to think about experimental approaches, then, approaches that seek to move out of the traditional boundaries that define critical thought, various and exhilarating and important as so many of them are, and to explore more options for thinking.

In this context I am heartened by many recent studies by ecology-inflected scholars who have begun to push against the boundaries of the academic book as traditionally defined. Anna Tsing's *Mushroom at the End of the World*, for example, tells a different anthropological story (of the matsutake mushroom) in a different form. In a moving passage, Tsing acknowledges that she wrestles with linearity but, at the same time, does not know how to think about social justice without the idea of progress. How does one articulate and work toward social and environmental justice goals, she wonders, without thinking about progress? She suggests that we require "new tools for noticing," among which she includes "interruption" (37); indeed, she organizes her book as "a riot of short chapters" that "tangle with and interrupt each other" (viii). Sylvia Wynter and Katherine McKittrick adopt a distinctive format—at once bold heading, conversation, and argument—to convey Wynter's theories in "Unparalleled Catastrophe for our Species?" Karen's Pinkus's *Fuel: A Speculative Dictionary* riffs creatively on Jules Verne and the energy humanities in a series of alphabetized "dictionary" entries. Donna Haraway, in *Staying With the Trouble*, appends a science fiction story to her study of climate change and what she calls the "Chthulucene." The edited collection by Anna Tsing et al., *Arts of Living on a Damaged Planet*, draws the reader's attention to the work's form and mediation: it is printed as two reversible books bound together as one, its print fluctuates between black and grey (the latter to indicate the idea of "ghosts"), and illustrations are innovatively placed throughout.[9] Another edited collection entitled *Textures of the Anthropocene: Grain Vapour Ray* by Katrin Klingan et al., similarly draws attention to its form through the paper stock for the different covers of its three volumes—rough, smooth, and shiny, respectively—and it explores a number of different visual strategies for delivering its material. Kath Weston's *Animate Planet* stops midsentence.[10] All of these works, and many others, invite us to see how form is always part of the story that is told. They also remind us that there is always another way of writing an academic book.

> ***Waypoints:*** *Central Park, designed by Frederick Law Olmsted and Calvert Vaux, is a large, rectangular, wooded area in the middle of Manhattan. The country in the city. On the morning of 9/11, I rode my bicycle through the park with my husband and our three-year-old daughter. We had just watched the Twin Towers fall. We had also watched the gaping hole left behind, blue sky and white smoke, unable to move or look away. When we entered the park there was a palpable shift in the air—the alarms of the fire trucks were more muted, the scent of autumn leaves more intense. Every hundred yards or so, people were clustered around transistor radios listening for news. Everybody wanted to know. We stopped and listened—You saw what? What? What? And then continued biking. I am returned to that day because we are in the midst of another crisis, again seeking news, listening with others, exchanging stories. I am returned to that day because then, on 11 September 2001, we were in the*

midst of the current climate crisis and I did not know about it (or, rather, I did, but it was like a small, hollow ball that had lost its track and rattled every so often from somewhere deep in the machine but I was never sure where it was or what it meant). I am returned to that day because I cannot bike away from the climate crisis, cannot find temporary solace and community in the park, and cannot tune into any clear station that will tell me what is happening and what I should do. But mainly I am returned to that day because of the 2019 release of a new docudrama miniseries, Ava DuVernay's When They See Us.

In 1989 Central Park catapulted to public attention when a white female—known as the Central Park jogger—was brutally raped and beaten. Five boys, known then as the Central Park Five, were arrested, detained, interviewed, and ultimately convicted for sentences of between five and fifteen years; in 2002 these convictions were overturned and the men were exonerated. DuVernay's miniseries revisits the story from the perspective of the wrongfully convicted boys. The docucrama is painful to watch as it starkly represents the ease with which a corrupt political system extinguishes the hopes and possibilities of these boys' lives.

On the night of their arrest, the boys were asked what they were doing in the park. They responded: "Whiling." "What's whiling?" the police detective asked. Not satisfied with the boys' answer, the detective constructed one herself: whiling was wilding. This "translation" has been taken as an example of racism (which it is), but I want to think about it, too, in relation to the force of words. As DuVernay notes in an interview, words are weaponized against Black people. For the boys, whiling means hanging out, whiling away the time. For the white detective, it means being wild. These opposing ideas come together and get defanged and naturalized with Thoreau and so many other "nature writers" who while away time in the wild. Not so for Black teenagers in Central Park. The Central Park case is a potent reminder that, yes, words are weaponized. But also: words are dense and alive with meanings that do not always align or that are violently misaligned. I want to think about what it means to mistranslate, to try to translate, to fail to translate, and also not to try at all. Words are weapons. They are the unrepresentable, unspeakable difference between whiling *and* wilding.[11]

My third rationale speaks more directly to my specific contribution in this book: the role of time and, in particular, interruption in climate change thinking. In this context, I have found Walter Benjamin's work both inspiring and provocative. While many critics, like Benjamin, challenge what he calls a universal, absolute, and empty time consonant with linear histories, very few do so with such fine attention to form, method, and mediation. This focus is inseparable from his desire to think of history as surcharged and, importantly, as unrepresentable moments puncturing, but by definition never realizing, the present. In other words, Benjamin challenges the often-unquestioned link between representation and knowledge, and he does so through a close consideration of the materiality—form, method, and mediation—of knowledge. His efforts to mount this challenge lead him to introduce many alternatives to teleological and linear form: montage, quotation, layering, constellation, superimposition, and interruption, among others.[12] His description of Bertolt Brecht's epic theatre "as a series of experiments" that generate "astonishment" ("Author as Producer," 235)[13] could also be a description of his own writing. That said, Benjamin's famously difficult texts may not seem the most obvious place to turn in my desire to broaden the audience for humanities-inflected approaches to climate change.

Nevertheless, I am not alone in finding Benjamin's work a productive source for climate change thinking. In particular, Benjamin's final essay, "Theses on the Philosophy of History," written in 1940 in the months leading up to his ill-fated attempt to seek asylum across the Spanish border, offers a disquieting meditation on representational and linear histories of progress that for many scholars have resonated with our differently imperiled period of climate precarity. Benjamin's essay captures, with uncanny precision, the sense of paralysis and propulsion by which many in the Global North are daily confronted in the context of climate change. In an oft-cited passage he writes:

> A Klee painting named "Angelus Novus" shows an angel looking as though he is about to move away from something he is fixedly contemplating. His eyes are staring, his mouth is open, his wings are spread. This is how one pictures the angel of history. His face is turned to the past. Where we perceive a chain of events, he sees one single catastrophe which keeps piling wreckage upon wreckage and hurls it in front of his feet. The angel would like to stay, awaken the dead, and make whole what has been smashed.

> But a storm is blowing from Paradise; it has got caught in his wings with such violence that the angel can no longer close them. This storm irresistibly propels him into the future to which his back is turned, while the pile of debris before him grows skyward. The storm is what we call progress. (257–58)

From Benjamin's interwar writing to our own period we are swept along, with a force at once violent and elemental, by a storm we, too, "call progress." It gives no indication of letting up, and many now discuss, with varying degrees of resistance or acceptance, a world "without us."[14] Faced with a differently bleak future, Benjamin picked up a pen and wrote. He wrote essays, he wrote books, he wrote letters, he wrote those documents now forever lost in his missing suitcase, and he wrote notes toward his unfinished *Arcades Project*. I can appreciate that turning to writing and thinking may seem a thin response to climate peril, but I also pin my hopes on the possibility that if we continue the conversation Benjamin began, we can begin to build better responses to the current and the coming "storm."

I am grateful that Benjamin gave us these images of the angel and the storm through which to filter and amend our own experience. I am grateful, too, that his comments have generated so many insightful responses, seeding a long afterlife of climate change conversations. One of the first essays to bring climate change into dialogue with the humanities, Dipesh Chakrabarty's pivotal "The Climate of History: Four Theses," for example, takes its impetus from Benjamin's essay with its "angel of history" and eighteen theses.[15] Michael Löwy's *Fire Alarm*, Andreas Malm's *Fossil Capital* and *The Progress of This Storm*, Ulrich Beck's *The Metamorphosis of the World*, Ian Baucom's *History 4° Celsius*, Priscilla Wald's "Science and Technology," and Bruno Latour's "An Attempt at a 'Compositionist Manifesto,'" all invoke Benjamin's "Theses" to sharpen their own climate change thinking.[16] Malm incorporates Benjamin's language as follows: "From the very start, at the very smallest scale—there emerged a pattern—some swept away by the storm we call progress, others sailing to their fortunes—subsequently magnified and iterated on progressively larger scales, until climate scientists discovered it in the biosphere as a whole, where the self-similar storm now spirals on. Every impact of climate change unfolds a fraction of that hitherto folded past" (*Fossil Capital*, 393). He uses this analysis as a springboard for at once despairing—climate change should be "the movement of movements"—and inviting, even now, the required action. The "question is—as so many have pointed out—whether [the climate change movement] can attain that status and amass a social power larger than the enemy's *in the little time that is left*" (394, emphasis in original). What Benjamin's work asks us to do, however, is step away from these sorts of configurations of time—what little time is left—and to inhabit time differently. To inhabit time in a way that multiplies it. To inhabit time, above all, in a way that makes *it* the movement of movements—in other words, the dialectical image or dialectics at a standstill—that animates a relation that cannot be pinned down or circumscribed and that opens the now (now-time or *Jetztzeit*) into a multitude of possibilities, doubled and redoubled, from which descriptions like Benjamin's the angel of history emerge. This approach is indelibly bound up with Benjamin's powerful and idiosyncratic treatment of the concept of interruption.

"Interruption," David Ferris writes, is "an essential starting point for reading Benjamin" (4). Benjamin's interest in the concept spans his career from his early work on the Romantics through his later more philosophical writing (in which Andrew Benjamin identifies a "more generalized sense of interruption" [105]), through the *Arcades Project* that exemplifies what we might think of as interrupted form. Interruption is also bound up with an obstacle critics almost always note when writing on Benjamin's work: its resistance to paraphrase.[17] Benjamin's work resists paraphrase because, like poetry, it conveys its meaning through its form. Any effort to offer a summary or commentary will be a distortion of something fundamental to Benjamin's contribution to thinking. And yet to write about Benjamin, one must summarize, and such objections to summary, while valid, can appear trivial. Is not every summary a necessary distortion that enables us to animate and build on the writing and to serve its vitality over time? Is it not enough to notice the form of the work, distil its meaning from that form, and illustrate how the meaning intercuts with the form when applicable? This is what scholars, myself included, tend to do. It is a valuable exercise. But *Climate Change, Interrupted* asks what happens when we pause, when we interrupt that practice, and attend to the form for a little longer. Consider Ferris's quotation with which I began this paragraph. To reinforce and validate my focus on interruption, I drew on an established scholar whose book is in a respected Cambridge series. But if we return to Ferris's claim, we can quickly see the limitations of the summary mode here. Benjamin, as Ferris reminds us later in his text, does not support "essential starting point[s]." Further, how could interruption be a starting point? To interrupt something, doesn't one have to have started already? Interruption—even as a concept—does not seem capable of being a starting point if by that one means, as Ferris seems to, the first point on a continuum.

But I agree with Ferris: interruption is integral to Benjamin's thinking. Ferris, moreover, is not alone in making this claim.[18] Close attention to interruption illuminates not only the reservations about interruptions as a starting point noted above but also, more fundamentally, the idea of always divided origins, in general.[19] As Ferris notes, the interruption introduces a pause that Benjamin, in his work on the Romantics, links with the breath and the body. Benjamin writes (in a passage cited by Ferris): "Tirelessly thought begins continually from new things, laboriously returning to the same object. This continual pausing for breath is the mode most proper to the process of contemplation" (*Origin of German Tragic Drama*, 28; cited in Ferris, 4). The pause enables the reflection that our digitally interruptive culture tends to compromise. For us, interruption often coincides with distraction. Whereas for Benjamin it is the opposite: a doubling down on thinking. If we're all distracted enough, the brute force of capitalism and neoliberalism can have their way.[20] In this sense, too, Auden's much-discussed "Poetry makes nothing happen" emphasizes how it makes *nothing* happen: it produces the pause.

If interruption is suggestive of a pause that prompts thinking, it is also, Andrew Benjamin argues, identified with the break wrought by modernity (104). Modernity, of course, is variously defined. Susan Buck-Morss, however, identifies 1850 as a pivotal turning point for Benjamin. It is in this period that the impress of new technologies is inescapably felt; it is this period, too, with the rise of industrial modernity, that many associate with the origins of the Anthropocene.[21] In 1851, London's Great Exhibition and the Crystal Palace that housed *it*—with its iron and glass construction and its promotion of English industrial advances—was an apt symbol for this shift. And 1851 was the year in which Henry Mayhew published *London Labour and the London Poor*, a brutal reverse mirror image of the Great

Exhibition documenting the lives of the urban poor made precarious and insecure by the very processes that made the Great Exhibition such a success. Mayhew's account, moreover, adopted an inventive, unprecedented form suggestive of print culture's potential to chart new avenues for thinking, a project to which I will turn in greater detail in "Post-time." Benjamin does not reference Mayhew, but both writers are sensitive to new modes of writing history that value the transient, the mobile, the discarded, and the humans and nonhumans who have not traditionally figured in the historical record. Narratives of progress, as Benjamin famously notes in "Theses," are narratives told from the point of view of the oppressor. We can interrupt that story to tell a different story. But Benjamin is not just calling for a different documentary account with different historical actors. Rather, he is calling for a different understanding of the documentary—or representation—altogether.

> **Waypoints:** *Meanwhile. Fires crest a hill. The sky is orange. This is what whiling looks like when wildfires wild. A colleague, learning of my interest in the Amazon, recommends Eduardo Kohn's* How Forests Think.[22] *I watch the fires on screens and I wonder what forests are thinking now. I wonder if the thinking is burnt out. Burn out: what happens when one fights fires, gets no sleep, makes no headway. What happens when there's not enough whiling. But also, burn out: what happens when there's not enough wilding.*

Interruption, etymologically, is a break in a continuous line. When Andrew Benjamin suggests that modernity marks an interruption, he invokes, accordingly, such a break (97). And, to be sure, the title of this book, *Climate Change, Interrupted*, gestures toward the possibility of interrupting our current climate change course. But, building on Benjamin's thinking, I also want to explore another, to my mind more potent, register of interruption. Decisive breaks are not the norm. Another form of interruption, however, occurs when old forms of understanding are no longer adequate to current conditions. In such contexts, one interpretation is overlaid on another and there is a disconnect; the two versions cannot be reconciled and yet they both coexist. Consider the so-called decisive break of 9/11, for example. In retrospect, that day seems to mark a change, a before and an after. But, at the time, as we tried to make sense of what was happening with the tools at hand, those tools were inadequate. Instead, there was a gaping hole. New words like "terrorism" and "homeland security" and "alert levels" entered our vocabularies. At some point those new words become the new conceptual frameworks through which we understand our world. But those adjustments take time, and they happen incompletely. A "break," in other words, takes time to be felt, to penetrate the social field. And as we adjust, there is another form of interruption—the gap between incommensurate, dissonant interpretations—that obtains. It is this form of interruption that Benjamin's work also helps me to understand, a form of interruption that is attuned, to borrow Benjamin's terminology, to the flow and arrest of thought.

This understanding of interruption strikes me as useful in the interwar years when there were at once decisive breaks and the ongoingness of daily life. It also strikes me as important in our own period of climate change upheaval. Precisely because climate change is not a decisive break, it brings into relief the overlaying or superimposing of discordant views that also apply across quicker timeframes, in periods of more decisive transition. We witness all around us politicians and others promoting contradictory positions simultaneously. In Canada, as I am writing, protests are taking place across the country in opposition to the advancing of a pipeline through Indigenous Wet'suwet'en territory. Canadian politicians recognize and promote the seriousness of climate change, and they know we must act. At the same time, they give the green light to a bill that contradicts this view entirely. All of us in the Global North daily enact contradictions on a smaller scale; we know the reality of climate change and yet we get in our cars, heat our houses, make our vacation plans. My point is not that we are wrong to do so, or even that the government is wrong to support the pipeline (although I think it is). My point is that we all occupy places riddled by interruption in this sense. In some periods that interruption is more muted than in others. It makes sense, however, to learn to recognize when we reach for forms of thought because they are familiar even when they are inadequate to lived conditions.[23]

In addition to understanding interruption as a break and (more pertinently, I am arguing) as a gap between incommensurable positions, several recent critics have, as I am doing here, linked interruption specifically to our current moment of climate precarity. Roy Scranton, for example, borrows Peter Sloterjik's understanding of philosophy as interruption and relates it to climate change. He cites Sloterjik as follows: the "subject can be an interrupter, not merely a channel that allows thematic epidemics and waves of excitation to flow through it. The classics express this with the term 'pondering'" (86). Scranton glosses this point as follows: "Sloterjik compares the conception of political function as collective vibration to a philosophical function of interruption. As opposed to disruption, which shocks a system and breaks wholes into pieces, interruption suspends continuous processes. It's not smashing, but sitting with. Not blockage, but reflection" (87). This emphasis on suspension and "sitting with" comes close to capturing the sense of occupying a *between* or gaping hole that the *inter* in interruption signifies.[24] Tsing, as noted above, similarly turns to the role of interruption in responding to climate change. Her response, however, is one that focuses on method and form. She writes: "To listen to and tell a rush of stories is a *method*." "A rush of stories," she continues, "cannot be neatly summed up. Its scales do not nest neatly; they draw attention to interrupting geographies and tempos. These interruptions elicit more stories. This is the rush of stories' power as a science" (*Mushroom*, 37).[25] Sloterjik, Scranton, and Tsing, in very different ways, use interruption to highlight a form of thinking that pauses, turns things upside down and inside out, unsettles us.

While Benjamin's approach to interruption could be said to embrace the different modalities I have outlined above—interruption as thinking, philosophy, and decisive break as well as interruption as the overlay of incommensurable positions—something more fine-tuned in his account has been relatively overlooked in the critics who develop his work for climate change studies. In an effort to understand Benjamin's pursuit of "pure language," Samuel Weber cites Benjamin citing Holderlin: "Tragical transport is namely actually empty, and the most untrammeled.—Thereby what develops in the rhythmic sequence of ideas wherein the transport presents itself is that which is called in prosody a caesura, the pure word, the counterrhythmic interruption, necessary, in order to meet the rush of ideas, at its height, so that not merely the change in ideas appears but the idea itself" (78). Weber glosses this now twice-removed passage as follows: "Pure language as the word that is without expression, pure word, is here designated as the caesura that, all of a sudden, stems the rush of ideas, arrests its flow, cuts against the grain of grammatical meaning" (78). This passage, in turn, recalls Benjamin's well-known comment on thinking (itself a rephrasing from another iteration in "Theses"): "Thinking involves both thoughts in motion and thoughts at rest. When thinking reaches a standstill in a constellation saturated with tensions, the dialectical image appears. This image is the caesura in the movement of thought" (*Arcades Project*, 67). The standstill, constellation, dialectical image, and caesura all invoke, in different ways, interruption. As Andrew Benjamin suggests, "Perhaps the most emphatic" treatment of interruption in Benjamin's work is "the 'caesura'" (98).[26] To caesura and these other words, Weber also adds "pure language" and the "pure word" as concepts which arrest the "rush of ideas." Pure language interrupts and suspends the sequential rhythm.

Pure language also raises questions about translation or, as Samuel Weber puts, translat*ability*. Weber alerts our attention to a suffix often used in Benjamin's writing: *-abilities*. As Weber notes, *-abilities* is a time word in German—"aka *verb*, that is inseparable from time insofar as it involves an ongoing, ever-unfinished, and unpredictable process" (7). In English this suffix often goes unnoticed. But once it has been noticed, it is impossible to overlook. It underscores the idea of unrealized possibility—something could be translated but it is *not yet* translated—and, in Weber's thoughtful commentary, the impossibility of realizing that possibility even when one tries. Paraphrasing itself is only a variation on translating. Quite often, for example, Weber cites a passage from Benjamin and then "translates" it into his own words. All literary scholars do this, as I noted above. We cite passages and then we elaborate on them. But Weber does so while respecting the movement and media of Benjamin's thought. I mention this point for two reasons. First, I worried about folding Benjamin's work into my study when I do not myself speak German. For a critic like Benjamin, whose thought is so forcefully embodied in his words, style, form, and structure, this ignorance struck me as an insurmountable shortcoming. Further, while Weber and other scholars point out mistranslations or alternative translations to prevailing versions, Benjamin himself often uses different words—the equivalent of the phrase *in other words*—to convey, seemingly, similar ideas. That is, he translates himself.

Second, if Benjamin's texts resist commentary and are *about* the impossibility of commentary, they do not resist "translatability." They are invested in and bound up with its possibility. They alert our attention again and again to mediation (although this is not Benjamin's word) and the ways in which it renders impossible documentary representation. As Weber puts it, "Translatability is the never realizable potential of a meaning and as such constitutes a *way*—a way of signifying—rather than a *what*" (92).[27] Pure language, he writes, happens when "ways of meaning, their distribution and relations, have priority over what is meant" (75). Pure language happens when translation hits a roadblock, when nothing happens, and a word sits there like a stone. In a trivial sense, one might say that all written language in foreign tongues are pure languages for us insofar as we see the language but we don't know what it means. Except that ways, distributions, and relations do signify in a pure language; indeed they come into sharp focus. Still, while it is an undeniable obstacle that I cannot read Benjamin in German, I take solace in the fact that my failure serves as a constant nagging reminder of the layers of meaning—the overlapping of the German word on the English typically captured in print by parenthetical notations, which are themselves a version of pure language—that comprise any work and the ways, distributions, relations it ignites. Benjamin's work foregrounds form—at the level of the word and the work's structure—to remind readers, through interruptions to the reading process that bring form into view, that dynamic and relational forms are *always* part of what and how one reads and thinks.

Consider again Benjamin's description of Klee's painting. It is thesis nine of eighteen theses and falls, accordingly, exactly in the middle of his essay. In doing so, its engagement with mediation, its in-between-ness or middleness, is foregrounded. And yet that in-between-ness is not a discreet item between two points but rather a distribution of relations, a movement, a hinge or fold from which the essay as a whole is indivisible. The description, moreover, follows a citation of a poem by Gershom Scholem that, based on my comments on translation above, is a different poem now that it is placed in this new context. And Benjamin's description, of course, is a description of an image. At the heart of these numbered theses, then, are a poem, an image, and a commentary. These four forms—numbered theses, poem, image, commentary—speak to the continued traction of these terms and ideas in thinking and dialogue. They are also examples of translatability. All of these things gesture to an afterlife, even if they will not reach it, and it is that afterlife that translation, at its best, captures.

They generate further afterlives to come as my own "afterlife" commentary here testifies.

> ***Waypoints:*** *Meanwhile. The fires in the Amazon are still burning, but their story has been displaced by the fires burning in Australia. And then those dim too, replaced by other news. I am planning my classes, and as more and more local bookstores close and libraries confront cutbacks, I find myself increasingly turning to Amazon to order books. They arrive the next day. My brother tells me that his books arrive the same day. For a teacher and reader, this feels like bliss. And then I learn that Amazon pays no taxes, despite their staggering profits, and what I know already sinks in: Amazon the bookstore's success mirrors Amazon the rain forest's destruction. I fuel that success, and in doing so I fuel the fire. I am one person, but still I vow not to order from Amazon again. I last less than a week. There are many things that I can tell myself—that systemic change is what we need and my small actions are just that, very small; that it is just one book (and one more); that I will redouble my energies in other directions—but still I cannot do this one thing. And yet, of course, that is not the point.*
>
> *Perhaps this is what is at stake in whiling and wilding. They both share a boundlessness. Time as loose and shaken out and running wild. They gesture toward the incalculable. And they ask us to act from that whiled, wild place.*

Writing and reading and thinking, for Benjamin, are about stopping and interrupting one's self, and this is also something I do as I read his work. I start and stop, start and stop. But if I interrupt myself as I read "Theses," "Theses" is also marked by two gaping interruptions that have a bearing on my discussion thus far: Benjamin's Paralipomena to the essay and Paul Klee's drawing of the angel. These items interrupt my reading because their absences are like holes in the essay signaling what is not there but could be. That is, neither the Paralipomena nor Klee's image are part of the essay proper, and yet they inform it. The former, moreover, includes one of Benjamin's most oft-cited comments on interruption, and the latter is now synonymous with Benjamin's writing. I will consider them in turn.

The Paralipomena is composed of, in the words of Benjamin's editors, "fragments [written] . . . in the course of composing 'On the Concept of History'" (*Selected Writings* IV, 408).[28] The English edition, the editor tells us, includes only a "selection" of these fragments. The one in which I am interested here speaks to revolution, thinking, and interruption:

> For the revolutionary thinker, the peculiar revolutionary chance offered by every historical moment gets its warrant from the political situation. But it is equally grounded, for this thinker, in the right of entry which the historical moment enjoys vis-à-vis a quite distinct chamber of the past, one which up to that point has been closed and locked. The entrance into this chamber coincides in a strict sense with political action, and it is by means of such entry that political action, however destructive, reveals itself as messianic. (Classless society is not the final goal of historical progress but its frequently miscarried, ultimately [*endlich*] achieved interruption.) (*Selected Writings* IV, 402)[29]

It is labeled Thesis XVIIa and falls, accordingly, between Thesis XVII, in which Benjamin—as he has been doing throughout the essay from different angles—again distinguishes between historicism (Ranke's history "the way it really was" ["Theses," 255]) and materialistic historiography (involving "not only the flow of thoughts, but their arrest as well"[30] [262]), and Thesis XVIII, the essay's last thesis, in which Benjamin reminds his reader of the miniscule amount of time—"one-fifth of the last second of the last hour"—that humans occupy "in relation to the history of organic life on earth" (263).[31] It is thus envisioned in Benjamin's notes as the penultimate thesis, but it did not make it into the final version. Instead we read it *into* that in-between space, before the end, between Thesis XVII and Thesis XVIII. Most readers also likely read it *after* they have read the essay and only insert these words retrospectively and provisionally. They are like a shadow text, an angel's wing.

The last sentence of this passage is especially interesting to me: "(Classless society is not the final goal of historical progress but its frequently miscarried, ultimately [*endlich*] achieved interruption)" (*Selected Writings* IV, 402). What does it mean to achieve an interruption? What Benjamin seems to be envisioning here is a departure from teleological histories—bourgeois, Marxist, and otherwise—that gestures toward different modes of meaning. It strikes me as apt, accordingly, that this passage on interruption not only does not make it into the theses proper but also is in a

FIGURE 1. Paul Klee, *Angelus Novus*, 1920
©Estate of Paul Klee / SOCAN (2021)

parenthesis. One of the definitions of a parenthesis is "interlude or interval," and, etymologically, it means "to place beside." Not in front. Not at the end. *Beside.* The phrase "*in parenthesis*," moreover, means to offer a "digression or afterthought"; in an essay comprising discrete theses by a writer who promotes a method of digression, this attention to rhetorical form becomes part of what the essay imparts. Like a Buddhist koan, one could dedicate years to meditating on this passage, and the remainder of this book could be devoted to elaborating its many possible meanings. Instead, however, I simply want to point to the ways in which interruption is bound up with these many possibilities and is activated by the form Benjamin chooses.

The second piece that interrupts my reading of Benjamin's essay is the "missing" image of Klee's drawing. Here, too, the positioning of the absent image is important. In Thesis VII Benjamin introduces the idea of "the genuine historical image as it flares up briefly"; describes historicism as a history of "cultural treasures" that not only ignores the history of the oppressed and what falls by the wayside but also is heedless of the force of the image; and includes the haunting phrase, "There is no document of civilization that is not at the same time a document of barbarism" ("Theses," 256–57). In Thesis VIII Benjamin introduces the state of emergency in which he lives (fascism and the onset of war) and calls for bringing about "a real state of emergency" (257), presumably something like the interruption to which he refers in the Paralipomena passage discussed above and on which I will elaborate in "Alarming." And then we have: a poem; a description of a painting. What are they doing here? Thesis IX turns a "cultural treasure"—Klee's image—to use for a revolutionary history that reorganizes temporal understandings. Unlike the Paralipomena passage, it is unclear where the image might be inserted. Should it be before Thesis IX? After it? In between the citation of the poem and the text? It seems to hover, or vibrate, in all of these spaces.

We can understand Benjamin's description of the image here as a "translation" in the "translatability" sense that Weber articulates above. A quick glance at it will illustrate, however, the degree to which the drawing does not correspond in any direct way to the words Benjamin records. What do you see? An angel of history, wreckage piled at its feet, looking at the past, his back to the future, a storm in progress? Or something else? Benjamin's description "translates" the image in a manner that tells us a great deal about his attention to the form, montage, and structure that he values in any interpretation; it attempts not the "inessential content" ("Theses," 253), to which Benjamin refers in the "The Task of the Translator," but that glimpse of translatability, of potential, of possibility. Climate change critics, as I have noted above, find this image, as Benjamin imagines it, a potent catalyst for climate change thinking. It radically departs from the more familiar images of the climate crisis: icebergs, polar bears, forest fires. These documentary images are intended to communicate information, create compassion, and compel action. But documentary images like these are demonstrably limited in what they accomplish, as several critics have illustrated and as I will elaborate in more detail in "Post-time."[32] Benjamin's angel of history offers a different way of thinking about the image by considering it in relation to temporality, history, and translatability. In this account, history is not a narrative told by the powerful that moves chronologically through time but rather a condensed materiality—more like a star or a fossil—that provokes and unsettles.

Benjamin's elaboration of Klee's image has not only been of interest to scholars. In his novel *10:04* Ben Lerner also references Klee's painting. The novel is set in New York City in the period between Hurricane Irene in 2011 and Hurricane Sandy in 2012. His novel grapples with climate change not by imagining an apocalyptic future but rather, like Benjamin, by compressing, condensing, and turning up time. Lerner includes Klee's image without commentary in his novel (at least without direct commentary) whereas Benjamin's essay includes commentary on the image without the image. In Lerner's novel, moreover, instead of having our back to the future like the angel (and also the movie *Back to the Future* that Lerner weaves through his novel), we turn around and around. To make this point, Lerner takes a passage from Walt Whitman's 1856 "Crossing Brooklyn Ferry" and reproduces it, save one word, as the final line of the book: "I know it is hard to understand / I am with you, and I know how it is" (240). This line is about union, community, returning, and creating through words a collective co-written bond to confront whatever faces us. What makes this line relevant to Klee's image and Benjamin, however, is that Whitman deletes it from the *Leaves of Grass* edition in which the final version of the poem is printed. Lerner, then, resurrects it and brings it back to the future, a potent reminder and harbinger, words as a holding pattern.

In *The Snows of Venice*, a collaboration with Alexander Kluge, Lerner returns to Klee's image and the "recent discovery" that it is "mounted on a print of none other than Martin Luther." This "secret hiding in plain sight," Lerner continues in a conversation with Kluge, is "remarkable . . . for a number of reasons":

That the "Angel of History," so long a symbol of left Jewish messianism, is mounted on top of Luther. That it places Benjamin's theses and Luther's theses in new relation. It's also a story about reproduction because part of the reason the print went undiscovered is that the borders of the Klee are normally cropped when the image is reproduced. Given Benjamin's thinking about reproducibility, this is a striking irony—that we haven't really seen the image in reproductions, that only a person who is physically present can see the traces of the engraving of Luther's cape. (81–82)

I agree with Lerner: this is a remarkable observation. For me, it was also a sobering reminder of mediation. Sobering because I had read Benjamin's essay many times and had discussed its mediations via its debt to earlier versions of the thesis form (Luther's theses central among them), and I had used Benjamin's interpretation of Klee's drawing as a commentary on the limits of representation *but* I had not thought of the materiality of the drawing itself. In other words, even when I was looking for mediation, I looked right past it. This discovery, then, also serves as a lovely materialization of the layering of meaning to which Benjamin refers in so much of his writing.[33]

. . .

Climate Change, Interrupted explores the ways in which the discordant temporalities specific to climate change pose a challenge to prevailing responses to crises. It does so by returning to the period of industrial modernity in which linear time took decisive hold and considering different approaches to temporality, now less often recognized, that emerged in the nineteenth century and resonate again in our own.[34] Each of the three iterations of these Beginnings also return to "Theses on the Philosophy of History"'s relevance to climate change in our own time.[35] The next Beginning, "Post-time," focuses on the relatively noninterruptive forms of realism and documentary representation. What happens to *forms*, this chapter asks, when new social actors and events are introduced? And how does changing the form relate to changing the time? To respond to these questions, this chapter develops and adapts George Eliot's term, "post-time," for climate change studies in relation to Henry Mayhew's *London Labour and the London Poor* (1849–51) and Richard Mosse's *Incoming* and *Heat Maps* (2017). The last Beginning, "Alarming," turns, by contrast, to an extreme of interruptive form, the rhetoric of warning. It takes Greta Thunberg's warning "Our house is on fire" as a point of departure for thinking about the rhetoric of warning and emergency in climate change discourse. If "Post-time" suggests a revision of our thinking about temporality and documentary representations in climate change advocacy, "Alarming" suggests a revision of our thinking about immediacy and the rhetoric of warning and alarm.

While the three Beginnings present their arguments in a relatively traditional academic format, the chapters that form the body of the book offer four different experiments in academic form. The first chapter, "Layering," is composed of six bands that imperfectly reproduce the geological stratification that has given us insight into deep time. The bands address Percy Shelley's "last days" on the Italian coast before his tragic boating accident, Jacques Derrida's "Living On / Borderlines," Virginia Woolf's reflections on the ellipsis in relation to interruption, and geology's stratigraphical reading practices, to tell familiar stories against the grain. The second chapter, "In the Idiom of the Self-Help Guide," develops the form of the self-help guide against itself. It toggles between the nineteenth century and our own period to harness Eliot's reflections on procrastination in *Middlemarch* for climate action.

The third chapter, "Found Questions," is a "found chapter" composed of questions excised from their contexts and put into conversation with other questions. Both Mayhew and Benjamin explore the resonance of the discarded or overlooked item for historical studies; this chapter applies that practice to questions by collecting questions from a range of texts and putting them into relation. The last chapter, "FrankenClimate," returns to the Shelleys, to consider Mary Shelley's *Frankenstein* in relation to unfinishedness, temporality, and interruption. Written in a series of nested frames modeled after Shelley's structuring of *Frankenstein*, it illustrates how frames often subdue the disorder that subtends any attempt to confront issues for which existing conceptual frameworks are inadequate. Figuring the Creature at once as climate change, a monster, and a blank, this chapter demonstrates what happens when we frame and hold the frame open at once. In short, it highlights a thematics of interruption and incompletion *and* interrupts traditional novelistic form to change the time and, accordingly, open new possibilities for thought and action.

Overall, *Climate Change, Interrupted* suggests a response to climate change that focuses less on representation, visibility, and linear narratives, and more on interruption, mediation, and thinking. It argues that changing the form is one way to change the time and that to change the time is to reorient the climate change idea. The multiple and layered conceptions of time that climate change itself registers, as well as the hard deadlines it delivers, offer alternatives to progress narratives and invite new possibilities. These new possibilities do not preclude progress, but they are not wedded to it. They ask us to recognize the important work that all stories do—fictional and otherwise—to chart new paths for thinking and imagining. In relation to the climate change idea that I am tracing here, these stories, moreover, are a collaboration in which we all participate whether we give our assent or not, contribute our voice or not, change the narrative or not, pass on a story or not, participate in a climate action or not. In other words, this is a story that is still being told, and its stakes, like other stories before it, will in turn shape the way we live and understand our lives.

I can write *about* interruption but, importantly, I cannot represent an interruption. As Benjamin illustrates, however, writing and thinking can facilitate interruptions, pauses, breath. Ferris describes Benjamin's work as advocating the "continually renewed beginning" (8).[36] Andrew Benjamin similarly notes that Benjamin's work opens a "field of infinite deferral" (99), and Buck-Morss emphasizes its insistence on beginnings (290). While Benjamin imagined starting a journal entitled *Angelus Novus* after Klee's painting, he never did so. And yet that journal perhaps begins again and again in different ways. When he crossed the Pyrenees, Benjamin carried a suitcase heavy with his most precious writings clutched to his chest. Every writer can identify with that sense of loss—not of the material item perhaps, but of its potential. I imagine the many diverse writings and conversations that Benjamin's comments on Klee's painting have generated as a kind of afterlife. I like to think of them as jarring us out of our torpor and prompting a different sort of action. I like to think of them as interrupting our response to climate change and provoking the *novus* of the painting, the angel newly visioned. I like to think of us pausing, seeing the wreckage, catching our breath, and facing the flames.

Waypoints: *On a warm spring day in 2018 the writer Nathan Englander went to Prospect Park in Brooklyn with his family. On that same day, he writes, David Buckel, a civil rights lawyer and environmentalist, also headed to the park. When Buckel got there, he doused himself in fossil fuels and lit a match to protest our nonresponse to climate change. Englander was struck by the fact that no one really noticed or registered this protest. He closes his article as follows: "Sad as Mr. Buckel's death is, as uninspiring as it should be to others, if he set himself aflame to send a message, and it's impossible to unburn him, and too late to direct his energies another way, the least we can do is spread the word."*

o

The point is one degree.

About Time: Second Beginning

Post-time

In *Adam Bede* (1859), George Eliot's narrator describes what many Victorians referred to as an "age of transition" by way of the shift from agricultural time to what he calls "post-time" (557). Post-time is time measured by the "periodicity of sensations" that the postal service delivers, an experience of time that, in the narrator's view, redefines leisure. He compares this shift to other lost forms that are part of the transition to what we might now call industrial modernity.[1] He fondly recalls "bygone years . . . gone . . . gone . . . gone," in which one spent one's leisure time reading the newspaper, by contrast to his own period where many different media compete for attention. Here is the passage in its entirety:

> Leisure is gone—gone where the spinning wheels are gone, and the pack-horses, and the slow wagons, and the pedlars who brought bargains to the door on sunny afternoons. Ingenious philosophers tell you, perhaps, that the great work of the steam-engine is to create leisure for mankind. Do not believe them: it only creates a vacuum for eager thought to rush into. Even idleness is eager now—eager for amusement: prone to excursion trains, art-museums, periodical literature, and exciting novels; prone even to scientific theorizing, and cursory peeps through microscopes. Old Leisure was quite a different personage: he only read one newspaper . . . and was free from that periodicity of sensations which we call post-time. (*Adam Bede*, chap 52).

Lost labor, lost leisure, and lost time. Counter to the idea that new technologies would usher in greater time for leisure, the narrator suggests instead that these technologies change the fabric of time itself. Instead of the rhythmic, cyclical, slow time of leisure (allegorized in the "spinning wheel") one has periodicity, eagerness, and an accelerated tempo so new and pronounced it requires its own term, *post-time*.[2] Eliot's passage, then, is invested in capturing the contrast between an earlier slow-time register of temporality experienced by the characters about whom the narrator writes and a new post-time dimension of temporality experienced by the narrator and the reader.

As I was working on this project, I kept coming back to Eliot's passage. I did so because, as beautifully as it defines a new media time that is also relevant to our own moment—our own "now" with its post-time defined by myriad digital posts—it also seemed to signal something broader about time's multiple scales, its materiality, and its mediation. I wondered if *post-time* might offer a way of routing the temporal challenges posed by the representation of climate change through a single term, flexibly understood, that both encompassed them and spoke to representation itself. In this chapter I test this possibility by turning to two documentary works that, in different ways, speak to climate change. Henry Mayhew's *London Labour and the London Poor* (1851) documents industrial modernity and the London poor in the mid-nineteenth century, and Richard Mosse's *Incoming* and *Heat Maps* (2017) document the unfolding refugee crisis across Europe, the Middle East, and North Africa. My gambit here is that these two works, separated by over a century and a half, can be put into dialogue to comment on documentary realism's potential to intervene productively in the climate crisis debates and the multivalent temporalities of climate change.

Like all of my Beginnings, this Beginning is informed by a dimension of Benjamin's "Theses on the Philosophy of History." Here I focus on Benjamin's call "to brush history against the grain" (257): to revise overlooked or neglected histories and voices (the genre of history writing) and, through doing so, to revise the temporality of history itself. Like Benjamin's intervention in history writing, Mayhew and Mosse imagine not just an expansion of the stories that documentary (or history) tells but also a rethinking of what documentary (or historical) representation itself means. Their work suggests a shift away from the *accuracy* of representation to, as Mosse puts it, the *adequacy* of representation. As I will illustrate below, Mayhew and Mosse brush the blue book and the camera, respectively, against the grain to put pressure on documentary forms. In doing so, both works revise how documentary realism is understood in ways that also relate to time studies. That is, both works also speak to Agamben's revision of Marx: that "the original task of a genuine revolution . . . is never merely to 'change the world,' but also—and above all—to 'change time'" (*Infancy and History*, 99).

. . .

Why begin with a realist novel from the mid-nineteenth century to illustrate these points? In an influential essay on historical emplotment, Hayden White suggests that the transition from nineteenth-century realism to modernism is defined by a "profound sense of the incapacity of our sciences to *explain*," and our representations to describe, the unparalleled crises, from the Holocaust to ecological suicide, witnessed in twentieth-century modernity (52). The twentieth century introduced, he argues, "unimaginable, unthinkable, and unspeakable aspects" that put pressure on old representational models and demanded new forms. While the exceptionalism of twentieth-century atrocities may be questioned, the point White raises here is only intensified in our current period; what White refers to, passingly, in 1992 as "ecological suicide" has become, thirty years later, significant enough to merit its own word: *ecocide*.[3] White maintains, then, that nineteenth-century narrative—and Victorian realism in particular—with its linear, compacted story lines, was ill equipped to represent more recent atrocities. White neglects, however, the role of interruption and asynchronies in these narratives, a point to which I will return below.[4]

But White is right about ecocide: it does present a profound challenge to existing representational models. From Timothy Morton's introduction of the hyperobject through Rob Nixon's concept of slow violence, this point is now widely acknowledged.[5] Invested in a view that the representation of crises is one avenue through which to focus one's energies when seeking social or political reform, many of those who turn to documentary forms have been frustrated by their limited impact. From Al Gore's *An*

Inconvenient Truth (2006) through Edward Burtynsky et al.'s *Anthropocene: the Human Epoch* (2018), there is no doubt that documentary works have galvanized public attention, educated viewers, and energized a movement, but there is also no doubt that their calls for substantial climate action have not been realized. Documentary genres have often been understood to offer a transparent, relatively objective account of social issues in an effort to promote social or political change (Nichols 9). In this understanding, documentarians conspire to efface the form in an effort to spotlight the documentary exposé. Documentary form, as a result, has often been overlooked. Recent studies, however, have brought form to the fore. The editors of *Remaking*, for example, describe "documentary as a pliable, improvisational form" subtended by self-consciously situated documentarians attuned to the pitfalls and possibilities of their genre (Blair et al., 6). Instead of documentary's "promise of immediacy and exactitude" (2), moreover, they consider how documentary realism can be evaluated "on a longer scale" and through "alternative timelines" (7). This chapter, too, puts its emphasis on documentary's engagement with the "alternative timelines" that Blair et al. support. But it departs from their focus on the "longer scale." The documentary works I address here make mediation legible; in their multivoiced delivery and departure from linearity, they keep the gaping hole open, so to speak, and maintain precarity. New forms are required for new topics and, with them, new temporal modes come into play. In Mayhew and Mosse, these new temporal modes emerge through interruption. They invite one to ask how a genre that foregrounds the documentation of reality points at once to the limits of representation in confronting climate change and to its possibilities: what it can do.

These possibilities have to do with form and time. They emerged most vividly for me, as noted above, not in documentary realism but in its generic cousin, the realist novel. Not only is *Adam Bede* itself considered a quintessential realist novel but its chapter "In Which the Story Pauses A Little" is considered one of the period's most astute commentaries on realism, variously called Eliot's "aesthetic manifesto" (Yeazell, 107), her "famous excursus" on realism (G. Levine 212), and "the locus classicus of Victorian realism" (Levine, *Serious Pleasures of Suspense*, 104).[6] This commentary on realism is also, tellingly, a "pause" or *interruption* in the novel's narrative chronology. Indeed, it may be one of the best known narrative interruptions in the nineteenth-century realist novel.[7] While realism, like documentary, is often defined in terms of its fidelity to reality and its seamless production of a realist illusion, Caroline Levine and others have more recently argued that this sort of "self-reflexive interlude" is not a departure from realism but rather a "staple of the realist experiment" (Levine, *Serious Pleasures of Suspense*, 104). In this interlude, Eliot's narrator, like Benjamin in his essay nearly a century later, calls for a more expansive realism, one that will include *more* ("old women scraping carrots with their work-worn hands," for example [224]), and to make this point, the form and structure of the story changes: it is interrupted. When Eliot introduces new social actors, her narrator opens an interruption or pause in the story and, in doing so, alters the form and tempo of the account. Not surprisingly, moreover, her narrator's later introduction of post-time also occurs in a narrative pause when the narrator brings readers abruptly into the present and invites them to think with him about temporality.

Let's return now to the discussion of post-time in *Adam Bede*. Post-time is bound up with "what was," what once existed and is now gone. That is, the force of the passage's "now" is inseparable from the impress and loss of the past. I found myself wondering if I kept returning to the term *post-time* because it superimposed or layered two times so deftly in that single word: *post*. For if post and post-time refers to the mail to reference both its delivery time and periodicity, post-time also takes its meaning, in this passage, from the *past* and past-time,[8] the time of leisure (and the time of whiling, to recall DuVernay's docudrama), to which it is contrasted. Indeed, the critical purchase of Eliot's passage is less the accelerated temporality figured by the post (that is assumed to be well understood) than the distinction between two different modulations of time that it marks and holds.[9] On the one hand, we can understand these modes as sensibilities: slow and fast; leisure and (industrialized) labor. On the other, we can understand them as linear markers of time's passing: past and present; then and now. As an interruption itself in a realist narrative, moreover, the passage invites us *both* to embrace the linear transition it describes—from past to present, leisure to eager, for example—*and* to hold two temporalities—past *and* present, leisure *and* eager—together at once. Post-time, then, may be understood not only in relation to an accelerated tempo to time (its narrow sense in this passage) but also in a more expansive sense: time as palimpsestic, interlaced, multidirectional.

The material mediation of time, moreover, is built into the definition of post-time. The materiality of the post and postal technology changes the experience of time. Clifford Siskin and William Warner identify the introduction of a "public postal system" in the seventeenth century as one of the four "cardinal mediations" that define the Enlightenment (12). By the mid-nineteenth century, postal reform was a topic of lively debate, the stakes of which involved, as Richard Menke puts it, "a wholesale reweaving of the social fabric" (33). This reweaving arguably contributed to the sense of "post-time" as Eliot's narrator defines it, a post-time that shapes the narrator's experience of time as materially mediated by postal delivery.[10] But it also points to the materiality of time in general: the new material relations of production ushered in by industrial modernity shape how time is understood, and that understanding itself takes material form from timepieces to postal delivery to schedules and so on.[11] The word *post*, moreover, carries its own robust materiality: the posts used to build houses (post and beam), signposts, doorposts, the posts used to define the beginning and ending of races. Finally, there is the hinge, the hyphen in post-time, that works as a reminder of the materiality of language built into the word itself.

Twenty-first-century readers, however, likely find the most resonant inflection of *post* in a meaning of the word that only emerges in the 1960s: post as what comes after.[12] If post-time as accelerated new media time (represented in Eliot's period by postal delivery) comes *after* time, it suggests a revision of the category of time itself.[13] Overlaid on the multiple temporalities outlined above, one has post-time as *after*-time, a different category of time altogether. After-time does not conform to any of our current comprehensions of time but seems to refer instead to some inchoate period *after* time that, even in my description here, the words I use and the ordering of the sentence defies. I can only gesture toward it. If post-time gives me a word for thinking about multiple and discordant temporalities together, the after-time inflection of post-time gives me a word for thinking about what I cannot do in language: that is, thinking outside of or after time. It captures at once the deep-time of geological thinking that both precedes the human and imagines a "future without us" *and* names the possibility of another temporal mode not yet imagined. That is, when *post* modifies time in general, it puts pressure on the models of meaning that produce these definitions in the first place.[14]

In sum, my first consideration of post-time focused on its meaning in the passage (the delivery of the post and the ways in which it throws into relief the contrast between two temporal modes that are linear, now and then, while also overlaying them), the second shifted to a focus on the ways in which time is always materially and historically mediated, and the third—post-time as after-time—considers both time beyond the human (deep-time and the "future without us") and the possibility of a different category of time altogether. In this book, I use the term *post-time* to encompass all of these definitions simultaneously; that is, I strain the work a single term can do. I am suggesting, however, that we consider linearity as a heuristic that is often useful but never detached from the other temporal modes I have outlined, even those in tension or contradiction with it, as Eliot's passage succinctly illustrates.

Post-time, then, is not meant to dismiss the temporal categories that precede it. On the contrary, I want it to underscore at once a palimpsestic temporality (that includes those categories but also rethinks them) and the role of culture and media in our comprehensions of time. When DuVernay attends to *whiling*, for example, as a Black cultural practice obtaining at a certain time and place that had not yet penetrated the broader culture, she is working within the category of post-time as I define it here. When Eliot attends to post-time to lament a lost sense of leisure time, she is doing something similar. In both cases, DuVernay and Eliot are using a term—*whiling* and *post-time*, respectively—to explore what makes a sense of time vital and potent for the lived culture in which it is developed and used. As I noted earlier, a failure to attend to this lived culture in the case of the Central Park Exonerated Five had legal consequences: Black youth were wrongfully convicted for a crime they did not commit.

The wrongful conviction, one could argue, followed, in part, from translating *whiling* as *wilding*. This is an example that makes mediation vivid. The mistranslation draws our attention to the materiality of the word that is bound up with its broader material mediations (postal delivery, periodical literature, books, and digital posts among them). I preserve the hyphen in post-time as a material reminder of the way in which all words are themselves folded layerings of time, hinged to, and in dynamic relation with, their pasts, in an always vibratory polyphonic materiality even when it is not named as such. Both Mayhew's and Mosse's works demonstrate the sort of "temporal polyphony" that, as Anna Tsing notes, "progress stories" often obscure (*Mushroom*, viii). Post-time brings that temporal polyphony into focus while also doing something that the invocations of multiple and discordant temporalities in climate change contexts do not always do: it weds temporal polyphony at once to its material mediations and its resonance as part of a lived culture. Adjacent to but not the same as afterlives, it also leans into forms of futural thinking without losing its tether to the multiple and overlapping temporalities of which ideas of futurity are only a part.

Mediation, by animating relations, animates time; it reminds us, as Eliot's passage also does, that any articulation of time will always be in relation to what preceded it and what follows. By unmooring claims to fixed definitions (of time) or objective representations (of the world), it opens possibilities even as it reminds us that all such possibilities are bound indelibly to their material conditions. At a time when climate change representations are doubly hampered by both an effort to "get it right" and an effort, linked to this point, to convey the multiple, overlapping, and expansive temporalities that climate change invokes, the term *post-time* at once disturbs chronological narratives that uphold progress and distributes temporal possibilities across multiple channels unconstrained by linear models. It offers, to cite Eve Sedgwick in the context of queerness, "an open mesh of possibilities" (7), inseparable from their mediations, that signal in turn multiple avenues for response. This is not to say that the linear model does not also obtain. It does. Indeed, it remains one of the best modes we have to make sense of our world and it, rightly, underpins many of our efforts to tell coherent stories. But these stories are only ever provisional and when we tell those linear stories without keeping post-time in play, we foreclose possibilities and, as Mary Mullen astutely argues, unnecessarily limit the ways in which we can imagine and live the future.[15]

Mosse and Mayhew, like Eliot in relation to realism, take that most representational of genres—documentary—and brush it against the grain. By considering how their work activates time in new ways and interrupts narratives of progress—how it changes the time—new possibilities for response emerge. Their focus on the *adequacy* rather than the *accuracy* of representations dissolves the divide between the representation and what is represented by folding temporality and mediation into the work itself. Post-time, then with its improvisational, interruptive poetics and practices, puts pressure on progress narratives to imagine instead, post-time narratives. Taken together, Mayhew and Mosse gesture toward not just an expanded repertoire of documentary stories but also a different understanding of documentary representation itself.

. . .

To treat Mayhew's mid-nineteenth-century account of the London urban poor as documentary is of course anachronistic. The documentary genre was not introduced until the early twentieth century with the rise of film. But it is precisely this sort of anachronism that I invite in this chapter. Mayhew's work anticipates the visual turn in its attention to illustrations and in its elaborate sensory descriptions. It also anticipates Benjamin's concept of the dialectical image as itself capitalizing on anachronism and intensely compacted temporalities.[16] When it was published, however, Mayhew's work was considered an unparalleled exposé of the struggles and stamina of the London poor.[17] First in London's *Morning Chronicle* (1849) and then in the four volumes of *London Labour and the London Poor* (1851, 1861), Mayhew sought "to publish the history of a people, from the lips of the people themselves . . . in their own 'unvarnished' language" (1:xv).[18] Today, disciplines from sociology to anthropology to ethnography vie to claim him as their own and continue to build on his groundbreaking documentary methods: oral histories transcribed to the page, jovial alliances with interview subjects, and an eclectic, transfixed attention to everything he encountered.

But if Mayhew's contribution to documentary representation and the comprehension of the geographies and lives of the London poor has been widely recognized, his contribution to a new comprehension of temporality has not. Indeed, the social, political, and cultural turbulence of mid-century London—let's call it post-time in Eliot's narrator's narrow sense—provoked at once a questioning of existing forms and an experimentation with new forms, a process that also disturbed and unsettled prevailing ideas of time.[19] This turbulence reinforces the aptness of the new word Eliot's narrator coins a decade later and affirms how, when social shifts occur, new vocabularies emerge to express them. Mayhew's work, for example, is squarely situated in that frenzy of accelerated temporal modes that Eliot's narrator laments. Mayhew describes many breathless evenings returning from a day of interviews for the *Morning Chronicle*, sitting in the back seat of a speeding horse-drawn cab, scrambling to write his concluding paragraphs to meet a printer's schedule. It was a thrilling rush of deadlines, speed, and submitting work "the ink of which was hardly dry," only to be holding the "fresh wet copy of the paper" in one's hands fifteen minutes later (cited in Anderson, 105).[20]

Mayhew's project also exemplifies post-time in the way that I have elaborated on it above. Before turning to the complex temporalities in Mayhew's work, however, I want to consider the genre of the blue book that he "brushes against the grain."[21] In his Preface, Mayhew writes that *London Labour* is "the first 'blue book' ever published in twopenny numbers" (1:xv). Blue books were introduced in the context of the growing interest in what Carlyle called the Condition of England problem. With the rise of industrial modernity and the loss of common lands following from the Enclosure Acts, the poor and working classes were increasingly perceived as a problem requiring study and solutions.[22] From the 1830s on, a series of governmental enquiries was initiated into topics like the conditions of women and children, the condition of housing, the condition of factory work, and so on. These reports produced volumes and volumes of evidence—from experts in the various fields and, importantly, from the poor and working classes themselves—that were compiled in large folio editions bound in sturdy blue paper (hence their name). In the mid-1830s it was deemed important to make this information available not only to governments deputized to act but also to the public itself. But as Roger Wallins and others note, it is unlikely that they were widely read in this form. Bessie Rayner Parkes describes taking one of these volumes from her parent's bookcase and being stunned by the record of suffering she read; but she also describes how difficult the format of the book was to manipulate and how small the print was on the densely packed pages (73–76).[23]

London Labour, by contrast, was inexpensive and personal, conducted by a "private individual" (1:xv) rather than commissioned by the government. Mayhew could accordingly privilege the voices of the poor in a way that that governmental reports could not. While government commissions gathered vast amounts of verbatim speech from the poor as evidence, this testimony was at once framed by the political questions that motivated their studies and shadowed by the volumes of commentary from government officials that surrounded them. Mayhew instead brushes the blue book against the grain. He takes this authorized format familiar to his period and inverts its emphasis. His entries often led with, rather than subordinated, the voices of the poor, and he made his studies widely available to an audience disinclined to surmount the obstacles to reading blue books.

Mayhew's volumes, then, were insistently accessible for the very reason he states in *London Labour*'s Preface: to "cause those who are in 'high places' . . . to bestir themselves to improve [the] condition [of the poor]" (1:xvi). He wants to change people's minds—indeed the very people writing those blue books—and prompt action. To do so, Mayhew used documentary reportage to demonstrate the dire and unsustainable conditions that the poor and working classes endured in midcentury London. And he gained a widespread readership: *London Labour* reached not only those in high places but also middle-class audiences and the poor and working classes about whom he wrote.[24] *London Labour* also animates the multitemporal and mediated mode I have outlined above in relation to post-time. Like Eliot's historical novel, Mayhew superimposes two temporal periods and implicitly encourages his readers to reflect on the implications of that superimposition. Unlike Eliot, his superimposition was not one of nostalgic overlay but historical record. *London Labour* compiles an account of lost labor and, with it, lost leisure, that, if not stretching back to 1799—the year in which Eliot's novel is set—seeks to offer a window into a world preceding the most severe transformations of industrial modernity. Mayhew offers an assiduous record of the upheaval produced when old labor practices were replaced by new technologies. The street folk he interviews repeatedly convey their bewilderment in response to changing conditions and the "extinct" jobs they leave in their wake. The narrative is replete with phrases like "a living doesn't tumble into a man's mouth, now a days" (1:47) and "Ah! twenty years ago, or better, live poultry was worth following" (1:126). His text's record of loss, in other words, is folded into its documentation of the present and its post-time tempos.

Instead of only offering a chronological record of decline, with its stories of extinct labor practices and current conditions, then, *London Labour* also produces the effect of a palimpsest, often with different temporal periods jostling for priority within a single sentence. This sense is also heightened by the nonchronological compilation of material in the final volumes. Mayhew, pursued by debt collectors, was forced to abandon the entire project in midsentence three-quarters of the way through volume 2, leaving subsequent editors to fill gaps and find content in other sources. While volumes 1 and 2 move more or less chronologically, volume 3 includes material that mainly pre-dates the writing of the other volumes, with many chapters reprinted from the *Morning Chronicle* and other sources.[25] Finally, in one instance—the first entry in volume 1—Mayhew writes controversially *about* chronology and linear progress and, here too, undercuts the account he delivers.

This first entry presents a chronological narrative of historical progress—the movement from what Mayhew calls barbarism to civilization, nomadic to settled societies, and vagabond to citizen (1:1). But in his subsequent entries Mayhew jettisons the developmental account all together. On the one hand, he realizes that those "nomadic" and "uncivilized" tribes live in London today: they are the London poor. He marvels over how apt the anthropological schematics is and expresses surprise that others have not noted the parallels he draws between the two. He often returns to this idea, reassuring his readers that if only they will support the London poor—usually through systemic reform—the poor, too, could advance from uncivilized wanderers to settled civilians. On the other hand, he quickly dispenses with the idea that the London poor correspond to "wandering tribes," noting that itinerancy only applies to a small fraction of them; he is entirely unbothered by any tension between the haphazard form he adopts to present his material and the settled life—presumably not only in the world but also on the page—he ostensibly values.[26] He tries to present schematizations and categories that offer some control over his material, but the material routinely overspills them, and their limits emerge so quickly that any effort to quell the abundance of the record yields simply to getting the material on the page. Not surprisingly, Mayhew's work is typically read nonchronologically. Apart from myself and the other editors of abridged editions of *London Labour*, I doubt very many people have read them in chronological order, from beginning to end.

London Labour also offers a case study in print mediation. The work is a panoply of divergent material. In addition to the interviews with street folk for which he is best known, Mayhew includes long sections of previously published material on London and poverty (sometimes cited and sometimes liberally lifted from other sources), extensive commentary on his sources, ecological studies, etymologies, extended studies of new labor forms (the docks, for example), long footnotes, and occasional tables and images. The wrappings of the part publications, moreover, printed letters from readers and, in doing so, provided an alternate format for readers to debate and discuss what they had read and for Mayhew to respond.[27] The uneven presentation of all of this material draws the reader's attention to the work's mediations (its form, its status as a built text, its intimacy with midcentury print production, its relations to its own materiality). These mediations are enhanced by local features of the text: print reproduced in mirror image so that *London Labour* requires a mirror to read; issues that break off midsentence only to be continued in the next installment; and the footnotes, tables, lists, and illustrations that interrupt the visual layout of the page.

The footnotes introduced in volume 2 are especially striking. Where volumes 1 and 3 include only a handful of short footnotes, volume 2 includes forty-five footnotes, several of which are extremely long. Many of these footnotes drill down into the history of words, unpacking their roots and providing usage examples.[28] A footnote on page 284, for example, explains that "pansherd" derives from "the Saxon *sceard*, which means a sheard, remnant, or fragment, and is from the verb *sceran*, signifying both to shear and to share or divide" (2:284). This idea of the *sceard* or shard returns in a later footnote (2:403) and underscores a larger principle of fragmentation and division by which Mayhew's volumes as a whole are marked (etymologies are also offered in the footnotes for "haunsed" and "back-flues" [206], "scab" [232], "hard-core" [281], "rubbish" [281], "*sanc*" [333], "*reredos*" [338], "querying" [369], and "sewer" [403]; these join the many other etymologies in the body of the text). While some of the footnotes indicate sources or additional information, the vast majority open up the text from within to introduce these etymologies, tangential ideas, and commentaries. The impression, especially in the longer footnotes, is a text overflowing its already divided boundaries.

Mayhew's account is responsive to the material he recorded and bears the impression of a writer and editor struggling to find his form. The interview transcriptions retain their subjects' dialect and idiomatic linguistic traits as well as obliquely register the interviewer, whose presence is conveyed by the questions—unrecorded—to which the subjects seem to respond (see Steedman and Herdman). What is striking, however, and missing in most abridged editions, is that these seemingly unedited transcriptions are often followed by a commentary from Mayhew himself that repeats, sometimes word for word, what the subject has just said.[29] It creates a strange echo effect as well as a kind of stuttering that jars the reader out of any sense of temporal immediacy and reminds us that these documents have been constructed—knit together, often with the loose threads showing—by the editor. These mediations are further amplified by the text's print history. Many of the articles that Mayhew wrote for the *Morning Chronicle* are remediated in *London Labour* just as *London Labour* itself went through many further remediations over the course of the following one hundred and fifty years in its afterlives as plays, songs, poems, and digitized editions among others (Schroeder). In short, then, Mayhew's work at once mobilizes post-time in Eliot's narrator's narrow sense as well as in my expanded sense; with its palimpsestic approach to the past and its attention to time's mediation through the material conditions of the text's production and the materiality of those words and texts, Mayhew's work is ever alert to the alternative temporalities it begins to bring into view even as it carries buoyantly on, heedless to its own innovations.

Post-time as after-time obtains here too. There is an aspect of his work—and, indeed, all works—that speaks to after-time as the text is reread and renewed by future readers. These readers read the work through their own frames of perception. What Mayhew could not know in his period, and what we still barely register today, was that Mayhew was also documenting the early days of a transition to a carbon economy; for the "age of transition" that defines the mid-nineteenth century is also a transition to carbon, and Mayhew's many pages dedicated to coal as well as to urban pollution are part of that story. Another part of that story, also still pertinent today, is how the poor bear an unequal burden of that transition.[30] To be sure, this is post-time as temporally palimpsestic, but it is also post-time as time projected, impossibly, into the future in the deep-time sense of what-we-cannot-yet-imagine. At this moment, as I write at my desk in Quebec, I cannot imagine what the future holds. One of the most grievous challenges of climate change is that it does not even conform to the patterns my imagination usually follows: one day following another, one year after the next. Not to mention the planetary boundaries it brings into view. Climate change is exponential and nonlinear in ways for which we have not yet begun adequately to account. So, too, Mayhew sitting in the back of the horse-drawn cab could not have known that I would one day retrieve his words on a computer powered by the carbon economy—the coal plants—that his surveys and interviews and descriptions begin to trace.

When we pass Mayhew's work—or any work—through the developing room of the present, a different picture emerges, one that was latent there all along. And returning to the work with that new image in focus also lets the work in question tell us more, speak in a different register, tune time to a different frequency. Indeed, as Herbert Tucker notes, events are not discretely bounded by temporal makers, but rather should be seen "from perspectives more acrobatically multidimensional" than the linearity of the timeline (n.p.). Post-time encompasses these acrobatically multidimensional perspectives in its attention not only to the sensational and punctuated time of midcentury England but also to the past-times against which it is defined, the mediations it highlights, and the afterlives it anticipates. Just as Mayhew turns the documentary work of the blue book against its intended purpose, so Mosse, as I illustrate below, turns the military camera against itself. By putting Mayhew's work into dialogue with Mosse's, I want to overlay temporal periods, as well as print and image, to multiply our modes of response and, as Mosse puts it, "speak sideways" about climate change.

. . .

I began to think about this chapter when, several years ago, Jesse Oak Taylor posed a question for a conference seminar: Is there a current cultural work related to climate change that is illuminated by reference to a cultural work from the nineteenth century?[31] Taylor invited his seminar attendees to think especially about frames for storytelling. How do frames of perception, in general, shape what and how we see? Richard Mosse's 2017 exhibits, *Incoming* and *Heat Maps*, in combination with Mayhew's *London Labour*, it seemed to me, spoke to this question. *Incoming* is a collaborative, multipart video that addresses war; mass migrations in Europe, the Middle East, and North Africa; the refugee crisis; and climate change; *Heat Maps* is a photography installation that documents the sprawling migrant camps where refugees wait. While these topics do not, at first glance, seem to bear any relation *London Labour*, both Mosse and Mayhew, in their different ways, seek to introduce to the public record social actors previously excluded. Or rather, they seek to tell the stories of these social actors—the mid-nineteenth-century London poor and twenty-first-century refugee, respectively—in ways that recognize their humanity. In doing so, however, the very frames for telling their stories—both in relation to documentary and to time—were unsettled and transformed. *How* they told these stories, how Mayhew and Mosse folded their distinct technologies of seeing into the story, accordingly, also commanded my attention. Unlike Jules Michelet's moving description of history as giving voice to the forgotten dead, their records not only make the dead speak—with all the resonances with post-time that phrase suggests—but also make the process of dying, the *incoming* of dying, the Anthropocene itself, speak. This is how the process of dying manifests, they tell us. This is how history unfolds: not as a progressive, uninterrupted line but as a multitemporal palimpsest. And these works, as noted above, do so in a way that is familiar to us from Benjamin: by rubbing their technologies against the grain.

The catalogue for *Incoming* is more like a box than a book.[32] It is heavy, square, and thick. I need both hands to pick it up. The black-and-white images are printed on unnumbered glossy pages bled to the margin. What one notices, then, upon opening the book, is the lack of a guiding frame for the images, a lack that encourages movement itself to be the frame as one turns from one image to another. This movement, of course, is also signaled in the gerund of the show's title. The climate refugees are incoming, the military are incoming, Mosse's photography crew are incoming—all are part of an unequal process of ongoing movement. And so too are we, Mosse's audience, incoming as we enter this book.

The first thirteen images present different perspectives on a rising (or setting) sun. Some of the images are abstract while others are more distinct. In some of the images, the sun is off-center while in others it is in the crease of the two-page spread; in some it is obscured by clouds, in others it is duplicated or smudged. And then three pages follow in quick succession of a plane, a missile, and an explosion in a city. Later there will be images of a moon. The pages that follow—and the video installation that the catalogue seeks to capture however imperfectly—are replete with images of the military, refugees, aid workers, doctors, children, and others. They are also filled with images of technology; not only the airplane and the missile, although these often recur, but also helicopters, boats, motorcycles, bicycles, propellers, flashlights, strings and cables, binoculars, iPhones, headphones, and guns. And if people are continuous with their technologies in these images—the hard hats, goggles, headphones, and belts laden with gear that change the outline of the human—so, too, do they remind us of the technology of the human in the reverence accorded the back of a man's head, the human handprint, the gymnastic poses, the open eyes. The sun itself is a technology that warms and illuminates the world, and it echoes the technology that makes these images possible in the first place: the heavy military-grade thermal camera that images only heat.

When I first looked at Mosse's photographs, however—both on my computer and in the catalogue—my response was not to notice this intermingling of technologies or to reflect on the commentary on climate change they raised. My first response was a sense of haunting. Whether I haunted the images and the people they represented or they haunted me, I wasn't sure. My first response, indeed, was to feel as if Mosse's images were, in Agamben's sense, changing the time. These images did not respect temporal rhythms I knew. Their refiguration of time seemed apt when I considered the camps that are the focus of *Heat Maps*. Here time unspools to a different beat as people wait—usually not for hours or days but rather years—to be granted visas to livable lives. In this context, it is not surprising that the *Incoming* catalogue closes with an excerpt from Agamben's *Homo Sacer* (as well as an essay by Mosse entitled "Transmigration of

Souls"). Agamben writes eloquently of the refugee camp as the signature of the modern (*Homo Sacer*, 174, 181). He builds on and revises Hannah Arendt's treatment of the Holocaust's death camps to reflect on the camp not as a state of exception but rather as a norm. In their context, the rule of law is suspended and a "zone of indistinction between outside and inside, exception and rule, licit and illicit, . . . subjective right and juridical protection" emerges (*Homo Sacer*, 170). But they are also a zone of waiting, a zone in which time wavers. They are, in the terms I have been discussing here, a zone of interruption.[33]

The sense of haunting that first struck me as I watched *Incoming* stemmed at least in part from the fact that its images are composed of heat imaged as light. That light, the white glow or shimmer in the image, in addition to producing a slightly off-kilter effect, lends the images an insubstantiality that jars with what the images so clearly convey: tools of war, suffering, distress. Documentaries themselves often use the rhetoric of revelation—exposé and "bringing to light"—in their descriptions. What Mosse does, however, is to invert these ideas and focus not on what is brought to light or exposed but rather on the mechanisms of exposé itself. He images the heat, the light, the modes through which exposé is possible in the first place. In doing so, he resists "the visual language of television documentary story telling" and creates instead a "new grammar" and "a new visual language" (Mosse, "Transmigration," n.p.). This new grammar, linked to post-time and interruption, resists representation and linearity to highlight instead movement, ephemerality, and the reality that stirs not in the stability of the representation but in its fluctuations and mediations.

The images are discomfiting. They convey my complicity with the camps' existence as I watch from a safe distance. My predictable sense of daily rhythms, my ability to plan, to schedule, to write this book to a deadline, are all woven into the temporality that defines my life and provides something like agency. But Mosse's images and Agamben's writing, taken together, also remind me that these camps to which I contribute each time I follow those daily rhythms are part of a larger condition of waiting in relation to climate change. For we are all in that zone of waiting, of temporal wavering, albeit cushioned in ways that the camps are not. We wait. And in some communities that can no longer turn away from the changing conditions of climate change—rising waters and temperatures, floods and fires—we run. The language of waiting informs all those communities that live, as Dionne Brand puts it, "like refugees," those communities in which basic rights are denied, invisible walls prevent movement, and a different grid of value and treatment obtains. The camp, then, is a place that dislocates place and time: it is everywhere, and time frays. You are in it even when you do not know you are in it. It is post-time as after-time, the before and after replaced with a pause, an interruption, in which time changes.

By making the technologies of seeing visible, Mosse illuminates not only places and people but also temporalities. By extending his commentary on the camps to the figurative camp that informs most conditions of living today in one way or another—by involving us all, in other words—he displaces and layers the camp in a way that is also part of the displacement and layering of time. The round disk of the sun that opens the catalogue, for example, is duplicated in many of the other images: in wheels, lenses, propellers, headphones, goggles, cogs, missile heads, coiled barbed-wire, and the moon. Mosse invites the intimacy between the sun and the camera he uses—the images he records, after all, are of heat, which reads visually as light. But the sun also informs our temporal rhythms; when the human impact on the environment changes the way the sun warms the earth, for example, it also changes the way the sun's rhythms dictate time. Crops arrive early or late and agricultural temporalities are disturbed, more powerful storms redefine established weather cycles, and dry conditions in some areas increase the wildfires that, as I write, obliterate the sun on the west coast of the United States and make noon feel, by many accounts, like the middle of the night. In his work, moreover, Mosse himself seeks to produce temporal conditions that conform more closely to post-time in my expanded sense than to the sequential chronology of standard documentary. The story of "migration is as old as humanity," he writes; it stretches from Homer's *Odyssey* through the contemporary refugee crisis to anticipate a future "when climate change radically displaces huge populations." This temporal sweep is, for Mosse, evocatively captured by the heavy military camera he uses. It lets him "simultaneously evoke three modes of storytelling: the mythic, the documentary, and the science fiction" ("Transmigration," n.p.). The camera's technology of seeing, in other words, condenses these different time periods in a way that enables viewers to sense them all "simultaneously" or, again, in terms of post-time.

But my comments thus far have elided the military purposes for which Mosse's camera was developed. "There is a new kind of camera," he writes, "that can detect thermal radiation." These military cameras, heavy and awkward to use, are capable of detecting body heat, Mosse continues, as far away as 30.3 kilometers and identifying an individual from 6.3 kilometers.[34] Because the camera captures the subject's body heat through thermal imaging, Mosse wondered if he could repurpose the technology of the camera to think about the "idea of heat, imaging heat" and, through doing so, "speak sideways" to climate change and the refugee crisis (*Incoming*, n.p.). He wanted, he said, "to use the technology against itself, to brush it against the grain . . . . to enter into its logic—the logic of proprietary government authorities—to foreground this technology of discipline and regulation, and to create a work of art that reveals it." As Benjamin figures history and Mayhew the blue book, so Mosse uses the technology of the camera against itself while, at the same time, remaining attentive to its purpose as a "weapon of war" ("Transmigration," n.p.).

The camera, moreover, is specifically aligned with time. It reminds us that time is a collaborative project in which material technologies play an indissoluble and dynamic role. The camera mediates the past (the *Odyssey* and what Mosse refers to as, following James Joyce's phrase in *Ulysses*, the "transmigration of souls"[35]) the present, and "the future," and it also manipulates the pacing of time. The new media technology of this camera highlights slowness in contrast to the accelerated media speeds of our current moment. Instead of filming in "real time" or (like most documentaries) jump-cut time, Mosse slows the image down from "60 frames per second . . . to 24 frames a second to give the material a less distracting, more cinematic feel" and to heighten details that one would tend to miss "in real time" ("Transmigration," n.p.).[36]

We may recall now that in *Adam Bede* George Eliot self-consciously sought a new form—we could call it a new technology for telling the story of midcentury realism—that would register the peasant at the table scraping carrots along with the many other dimensions of daily life typically considered outside the purview of novelistic representation. And Mayhew, a decade before her, sought a new form through which to capture the impact of industrial modernity on the London poor, to preserve lost forms of labor and to offer a fuller, if still imperfect, sense of the complexity of the lives of the poor, while petitioning for social and political reform. The idea that to expand one's range of representation, new forms have to be developed has been oft-noted, but for Eliot and Mayhew this process was also accompanied by a revision in the register of temporality—what I am calling an expanded post-time here. Like Benjamin, they rebelled, in small ways and large, against linearity and sought a realism—not Ranke's history or reality "the way it really was" (cited in Benjamin, "Theses," 255)—that opened up the range of possible representations *and* disrupted time by drawing attention to the technologies of seeing and the work's mediations. *Adam Bede* is also a novel keenly attuned to mediation—from the drop of ink with which the novel opens to the postal delivery that post-time invokes—and it introduces many small side commentaries by the narrator like the one on post-time with which I began. The pause—the interruption—opens the novel to inquiry. It breaks its chronological sequence and brings the reader abruptly into the present. In Mayhew, the account of the London poor is presented in a bewildering array of piecemeal temporal dislocations, repetitions, and discontinuities that interrupt and destabilize the story of progress with which his account opens.

These techniques find a parallel in Mosse's work. In his "Artist's Statement" for *Incoming* (2017) Mosse, like Hayden White, discusses the need for "new images" adequate "to the ambivalence of the story" of the mass displacement of human populations, a story that is not only about war and conflict but also about climate change.[37] Instead of pursuing accurate representations, accumulating information, and falling back on linear narrations, his work asks what happens when the adequacy of the representation and its mediation is the focus. In both Mayhew and Mosse, the medium is emphasized. In his pursuit of the image adequate to the story of the refugee crisis and climate change, for example, Mosse dialed up his medium rather than focusing only on *what* he was representing. Indeed, he often leads with his medium when he discusses his work. This does not mean that the medium is determinative—on the contrary, Mosse deeply appreciates the surprises that stem from intuitions and experimentation with new mediums—but that the work is indelibly bound up with the medium. "There are many different ways you can tell any given story," Mosse remarks, and the "medium . . . can operate like a prism to focus the tale and give it clarity." In response to Werner Herzog's pursuit of images "adequate" to current conditions, Mosse explains, "I understand . . . 'adequate' to mean a non-reductive form, one that encompasses the ambivalence of the story in question" ("Artist's"). An adequate story maintains tensions, animates them, lights them up, foregrounds rather than subdues mediations, and invites uncertainty, invites us to ask, What is this? An adequate story is "a way to tell these stories . . . that keeps the heat on them" ("Artist's Statement").

My first impression of Mosse's work, as noted above, was a sense of haunting. Like ghosts, the images shimmer, blur, and reflect. They can feel like harbingers from another time. They demand our attention and imply that we ignore them at our peril. The overlapping of temporal periods invoked by ghosts recalls the dialectical image, a concept that has been in the background of this chapter thus far. In the previous chapter I cited part of the following quotation from Benjamin:

> Thinking involves both thoughts in motion and thoughts at rest. When thinking reaches a standstill in a constellation saturated with tensions, the dialectical image appears. This image is the caesura in the movement of thought. Its locus is of course not arbitrary. In short it is to be found wherever the tension between dialectical oppositions is greatest. The dialectical image is, accordingly, the very object constructed in the materialist presentation of history. It is identical with the historical object; it justifies its being blasted out of the continuum of the historical process. (*Arcades*, 475)[38]

The dialectical image appears as a caesura, or interruption, in the movement of thought.[39] Consider not only the ghostly aura of *Incoming*—a point noted repeatedly by critics—but also its angelic incandescence. If ghosts are implied throughout, so too are angels. Sometimes the ghosts are like angels. But sometimes we also encounter images of angels, images that, for me, invoke the dialectical image of the angel of history itself. Mosse arguably uses the dialectical image as an interruption to blast open "the continuum of the historical process" that Benjamin introduces. He ratchets up temporal tensions. *Incoming* includes many images of flying things: planes, missiles, birds, kites. Sometimes these flying things are ominous or deadly and sometimes they are not. In an earlier work, *The Fall* (2009), Mosse included photographs of airplane crashes in remote locations. These, too, may be seen as dialectical images, images from earlier periods that now resonate in our own. They interrupt the technologies of the airplane and war as we are accustomed to seeing and interpreting them, and produce a slight faltering in our response. These photos of lost war planes, flying things that have fallen, at once collapse—or, perhaps better, constellate—temporal periods to produce a blasting open of the continuum of history and, in this way, also anticipate *Incoming* with its subtext of falling and its leitmotif of ghosts and angels. Mosse's images, in general, may be seen as interruptions in the field of vision that cannot be resolved or sutured over. They mess with temporality.

Like Eliot and Mayhew, then, Mosse's work speaks to the Anthropocene in the register of time and representation together. Donna Haraway, like Dionne Brand, reads the idea of the Anthropocene (which she renames the Chthulucene) as producing the conditions of the refugee in general: "The earth is full of refugees, human and not, without refuge." In this context, she encourages joining "forces to reconstitute refuges, to make possible partial and robust biological-cultural-political-technological recuperation and recomposition" ("Anthropocene," 81). For Jenny Odell this is a call not only for "habitat restoration in the traditional sense" but also for "restoring habitats of human thought" (181). The two are inseparably bound together. The dialectical image refuses continuous histories, and in this refusal, creates a refuge that is not a place of protection necessarily but a place of possibility. It does not resolve, but rather holds and makes signify, the conflicts that chronology irons over.[40]

To return now to Taylor's question: How is Mosse's work illuminated by Mayhew's *London Labour*? Certainly military-grade images, taken at a great distance, of refugees in foreign countries seem remote from Mayhew's very different, close-up and engaged, account of the lives and labor of the London poor. But *London Labour* can be read as an indirect record of the results of mass migration that becomes legible, belatedly, when we read now. The climate change story latent in Mayhew's work emerges when one considers the work through a new prism. These frames of perception, too, are bound up with time and history. By putting these two works from different historical periods into dialogue with each other, I not only saw another instance of the migration story to which Mosse refers—the mass migration from country to city that had taken place over the eighteenth and early nineteenth century in response to the Enclosure Acts as well as the 1840s migration from Ireland to England in response to the potato famine—but also the role that mediation and movement played in both accounts.[41] In short, Mayhew and Mosse's use of the blue book and the camera, respectively, against the grain, supports Benjamin's broader call to brush history against its grain.

Brushing their technologies of representation against the grain, moreover, allows them to confront representation's limits and, in my examples here, to animate temporality, flexibly and palimpsestically understood, to open up different possibilities for documentary. If documentary cannot fulfill its initial promise to offer objective representations that generate social change, it can—and often does—offer an opening up of time. As Franny Nudelman notes, a sense of urgency, if not emergency, is written into documentary genres that focus on social injustice.[42] But Mosse's use of the media technology to slow time down explores new ways of catalyzing viewers and prompting action. Consider the following image from *Incoming*:

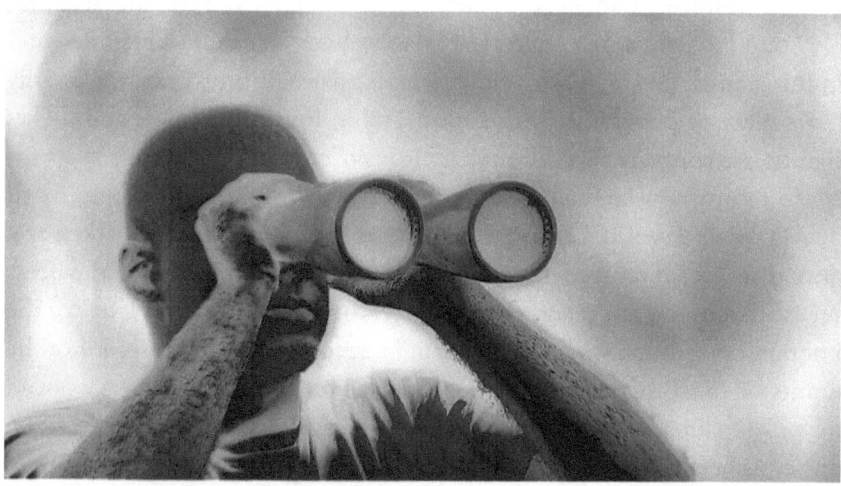

FIGURE 2. Still from *Incoming*, LXXXVI. Courtesy of the artist and Jack Shainman Gallery © Richard Mosse

This is not an image that pretends to be reality, but an image of the act of *seeing as mediated*.[43] Eliot dramatizes something like this in her reference to post-time when she recognizes that the accelerated time of the post changes the way that it is read. And Mayhew's work, while not self-consciously exploring mediation as Mosse's does, constantly reminds the reader of the ways in which its media forms—in newspapers, serials, and books among others—dictate how we read. Both Mayhew and Mosse's work, then, make time and mediation more legible—their work teaches us to read the mediations—by way of their focus on the technologies of representation in the context of layered temporal modes and "temporal polyphony," as they both also elaborate post-time as I have defined it here. In this version of documentary, the accuracy of representation no longer obtains and is, instead, replaced with the adequacy of representation. Both Mayhew and Mosse develop new technologies for seeing that point to the limits of documentary representations as traditionally understood. In doing so, instead of subduing the precarity of what they record, they introduce suspensions of time—pauses, interruptions, asynchronies—that make that precarity vibrate and register more intensely and help us rethink time as post-time: unsettled, relational, mediated, and "acrobatically multidimensional."

The weight of the *Incoming* catalogue reflects the weight of the military camera and, taken together, reinforce the weight of the topic they address. This weight is counterpoised, however, by the light in the images. It produces a frisson, a discordance, that remains unresolved and continues to disturb and unsettle me long after I have viewed these images. In the end, I am left thinking of Benjamin's angel of history, poised in midair, issuing his warning. I recall Mosse's images of paintings of angels: fallen from walls, halos and wings lifted in the hands of soldiers. The painted angels broken into pieces that people carry. These images of angels float in the middle of the photograph, decentered and decentered again. In one of the images, the warmth of a handprint is retained on the frame even after the hand has moved. This handprint echoes the many handprints that are included in *Incoming*, the trace of life, of warmth, that people leave after they touch something. *Incoming* is filled with hands: hands waving, reaching, helping, holding. Hands pressed to glass in a place of worship. They remind me of a moving passage from Robert Macfarlane's *Underland* that, in the context of ancient handprints on cave walls, unites hands across time: "I imagine laying my own palm precisely against the outline left by those unknown makers. I imagine, too, feeling a warm hand pressing through from within the cold rock, meeting mine fingertip to fingertip in open-handed encounter across time" (18).[44] Mosse's images remind me of what we leave behind when we touch a piece of cloth, throw it in the air, when we touch each other, when we touch war machines and water bottles, when we create the beauty of the angels and the blasts—a spray of light in the air—that destroy them. We throw our hands up. We wait. The handprint remains after we touch something and after we leave. For awhile.

. . .

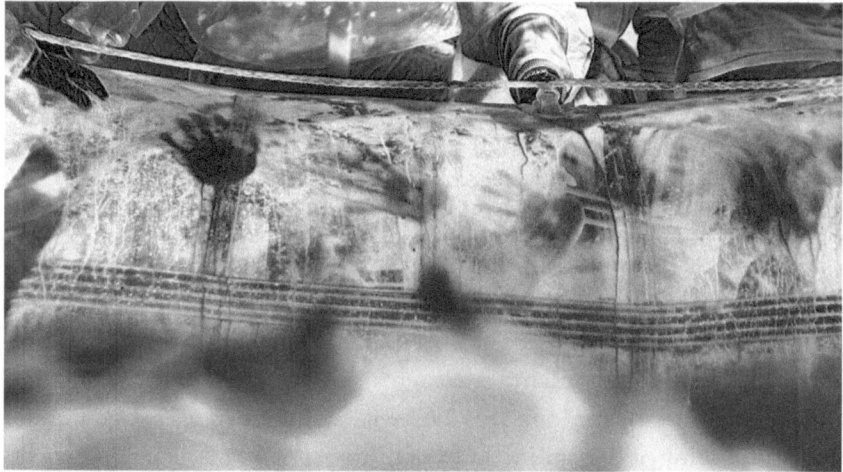

FIGURE 3. Still from *Incoming*, CI. Courtesy of the artist and Jack Shainman Gallery © Richard Mosse

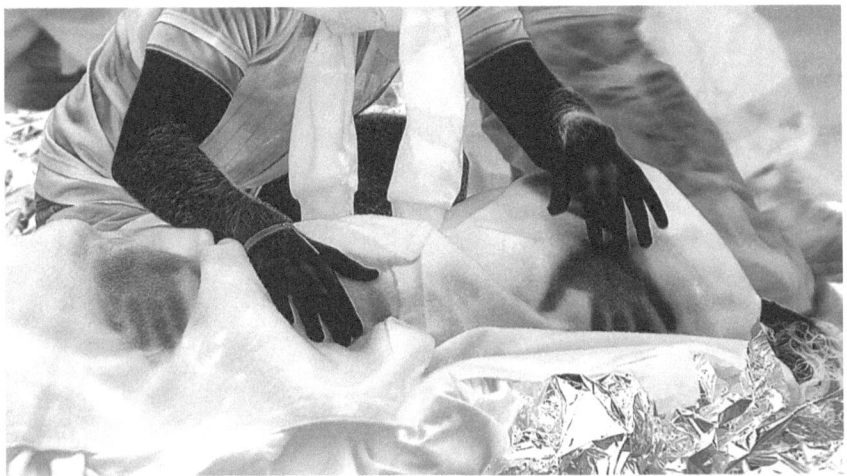

FIGURE 4. Still from *Incoming*, CII. Courtesy of the artist and Jack Shainman Gallery © Richard Mosse

FIGURE 5. Still from *Incoming*, CLXXXVII. Courtesy of the artist and Jack Shainman Gallery © Richard Mosse

About Time: Third Beginning

Alarming!

Our house is on fire

...

our house is on fire.

as if

I am here to say

...

I want you to act.

In a ten-minute speech, Greta Thunberg countered the considerable difficulties of representing climate change discussed in the previous chapter with a single startling image.[1] "Our house is on fire. I am here to say, our house is on fire," Thunberg warned the World Economic Forum in Davos, Switzerland, on January 25, 2019. In the conclusion to her speech, she returned to the metaphor of the house on fire: "I want you to act as if our house is on fire. Because it is." She mobilized the metaphor of the fire alarm and the rhetoric of warning to alert her audience to the danger posed by climate change. The danger is not idle. The house *is* on fire and the metaphor—"as if"—does not diminish the need for action now: *I am here to say. . . . I want you to act.* It was a call to attention. Her warning was immediately circulated and amplified across diverse media platforms around the world with a speed that *Adam Bede*'s narrator could not have imagined.

And yet, like documentary representations of the climate crisis, this riveting rhetoric of warning and appeal for immediate action had no appreciable impact on fossil fuel emissions. Indeed, emissions continued to rise. In the previous chapter I considered the ways in which realism and documentary support chronological sequence and yet, with the introduction of new social actors, seek to interrupt and open up those forms and, in doing so, produce new temporal modes. In this chapter, by contrast, I turn to that most interruptive of rhetorical devices, the warning. I focus on three recent critics—Naomi Klein, Bruno Latour, and Andreas Malm—who issue climate warnings themselves, offer anatomies of those warnings, and reflect on their impact. The fire alarm metaphor, after all, has a history, both the recent history we find in these critics and a longer history. Like Thunberg, Klein and Latour appeal to the burning house to emphasize at once the urgent need for immediate action and the global failure to execute that action. Malm deploys the metaphor of the speeding train to make a similar point. To be sure, each of these critics is wary of climate warnings (even as they continue to make them), and each invokes the meta-warning that warnings are not working. That said, *something* is happening in their warnings, and it is that *something* together with an alternative approach that I address here.

the burning house

the speeding train

something
something

Walter Benjamin's recourse to the same rhetoric of fire alarms and speeding trains in the interwar years offers insight into both why warnings to date have been unsuccessful and how, inflected differently, they might succeed. In the years leading up to the World War II, Benjamin presciently remarks that the continuous alarm *is* the catastrophe. What is needed in its place, he argues, is not the state of emergency that a constant ringing of the alarm registers but rather a "*real* state of emergency" spurred by interruption ("Theses," 257). Benjamin's call for interruption, variously defined, is tricky precisely because it is performed. Interruption is performed as revolution, but it is also performed, as I have noted, dialectically, in the materiality of language, rhetoric, and form. In the Global North, for example, the alarm often

the continuous alarm

the *real state of emergency*

upholds the temporal forms that contribute to the emergency it announces. One way to interrupt the logic of the alarm, then, is to take up, as Leland de la Durantaye puts it, "Benjamin's reflection on the pressing need for a radical reconceptualization of *time*" (115).

> *a radical reconceptualization*

Let's return to Thunberg's closing remarks cited above: "I want you to act as if our house is on fire. Because it is." Between the "as if" and the "it is," Thunberg captures the cognitive dissonance experienced by so many living in the Global North today. The climate emergency *feels* unreal ("as if") and yet most of us also know the emergency is real ("it is"). Climate warnings uphold a linear timeline that cannot comprehend this sort of dissonance. The warning interrupts the line, but instead of being an opening for confronting a crisis for which our existing conceptual categories are inadequate, it doubles down on the line and intensifies the emergency. Benjamin, by contrast, mobilizes the warning to hold together performative contradictions—the "as if" and "it is"—and to create a space for thinking "quite otherwise" (*Selected Writings*, 4:402).

> *to act as if*
> *as if*
>
> *as if . . . it is*
>
> *as if . . . it is*
> *quite otherwise*

1. BULL: ANATOMY OF THE ALARM AS WARNING

The OED defines the alarm as "an anxious awareness of danger." Its goal, the definition continues, is "to rouse to a sense of danger or emergency"; it is also a "warning of danger, *esp.* one intended to startle or rouse the previously unwary into action." Indeed "alarm" is derived, from the Middle French *à l'arme* as a call "to arms!" or a call to action. *I want you to act*, Thunberg says. The alarm puts us on the alert, it warns, it produces an anxious awareness of danger, and it encourages action in response to that danger. J. L. Austin's elaboration of the warning as a performative—a speech act that performs

> *Alarm:*
> • *call to arms*
> • *anxious awareness of danger*

what it says—in *How to Do Things with Words* helps us to unpack these different components of the alarm.

Austin famously distinguishes between *constatives* (descriptive language that answers to true/false questions) and *performatives* (language that makes something happen), only to dismantle the distinctions he so carefully draws as his book unfolds. While the promise is his best-known example of a performative, the warning is also an example to which he devotes considerable attention. A warning, he notes, only works in the context of a credible threat. If one were to say, "There's a bull in the field," but there were no bull in the field, then the sentence would not function as a warning. Similarly, if one were to say "There's a bull in a field" while looking at a painting of a bull in a field or out the window at a bull in a field, the bull would not pose a credible threat to the interlocuter and the speech act would not be a warning. But if one were standing in a field with a friend and one issued the same sentence, it could be understood as a warning. This phrase, then, depending on context, "may or may not be a warning" (32–33). If one says, "I warn you that the bull is about to charge" (55; cf. 74, 98, 141), the speech act's function as a warning is intensified. "I warn you that we have eight years to address climate change," is the climate warning's equivalent to the charging bull; one can immediately see in this example the time problem by which climate action is beset. Austin's illustration that constatives can be warnings ("There's a bull in the field" is, under certain conditions, at once a description and a warning), moreover, also holds in climate change language. Indeed, as Austin notes, the constative phrase, "There's a bull in the field," can be turned into a performative simply by adding an exclamation mark, "It's going to charge!" (74). Many descriptions, climate projections among them, do not require the exclamation mark

how to do things

warning as credible threat

there's a bull

I warn you ...

there's a bull

there's a charging bull!

to function as warnings, but the exclamation mark will intensify the warning. That said, as we know from the affinity between alarm and "alarmist," too many exclamation marks, figurative or actual, can blunt, or undercut entirely, the effectiveness of the warning.

alarmist

What makes an alarm successful? In Austin's terms, the alarm/warning is illocutionary (the speaking/writing itself performs the action) and the action it incites is perlocutionary (the consequential effect). Not all illocutionary acts are successful or, in Austin's words, "happily" performed. "I cannot be said to have warned an audience," Austin writes, "unless it hears what I say and takes what I say in a certain sense [what he calls "uptake"]" (116). Thus we may say "I tried to warn him but only succeeded in alarming him" (117). To be effective, a warning will alarm and that alarm will incite action. When the action does not happen, the warning fails. To extend this point to the bull example, the successful warning alerts one's interlocutor to the bull and, alarmed, she responds. The unsuccessful warning also alerts one to the bull but, alarmed, the interlocutor does not respond and is trampled instead. In the former case, the alarm produces action (the "call to arms") and in the latter the alarm produces anxiety (the "anxious awareness of danger").

happily performed

I say in a certain sense

I tried to warn him but

Climate warnings, however, depart from the bull warning that Austin outlines in at least three ways. First, unlike the bull, they call for collective action; if only I run (change my lightbulbs, recycle etc.), nothing happens. We all need to act in concert, a form of action inconceivable without either government directives or revolution. Second, again unlike the bull, the immediacy of the danger is not evident to all who hear the alarm. And third, the sequence from illocutionary to perlocutionary act, from alarm to action, is unclear. What does collective action even look like in this context? How long does one have before "the bull" charges? What action, exactly, is being called for?

To return again to Thunberg's climate warning that our house is on fire, we can probably agree that it at once fulfils many of the requirements for the credible warning while also introducing some difficulties. She made clear in this speech, and elsewhere, that her warning is based on reliable data. The problem is real, and she has the statistics and broader knowledge to support this point. She was clearly not joking, acting in a play, or asking a question. Further, she spoke in a respected world forum, and her voice was authorized and legitimated by that context. But at least two points detract from the credibility of her warning: she is young, and she couched the warning in a metaphor. The former point she turns to her advantage: it is precisely her youth and her need to speak that command our attention. The second point is slightly more complicated. The audience can look around and see that, despite rampant wildfires around the world, their own houses are not literally on fire.[2] Like the person who says that there's a bull in the field when there is no bull, this comment lends itself to being refuted. But Thunberg was counting on her audience to know what she meant, and she wanted to use the metaphor to heighten, rather than reduce, the force of her warning.

The metaphor of the "house on fire" indeed has many merits. Like others, Thunberg wants to communicate the urgency of the climate crisis by making it immediate. The house on fire reminds us that the Earth is our home at the same time as it brings the climate threat home: it is *like* the intimate space of the place in which we live. It also animates the specificity of the threat: the world is *warming*. It is getting *hotter*. When things get hotter, there are fires. And it is an urgent call: it is happening now. Finally, it articulates a clear action plan: wake up, put out the fire.

reliable data

the Earth is our home

it is warming
it is hot
it is on fire

> *we are doing things with words*
>
> *what more can we do?*
>
> *wake up and put out the fire*

And yet despite the credibility of the warning and the apparent aptness of the metaphor with respect to climate change, the parallel it draws breaks down: as noted above, the climate warning is not a call to a single person but to nations and corporations, and it is not a call for a simple action (evading a bull, for example) but for an unspecified collective action. Further, its reliance on chronological linear time poses difficulties when the topic addressed—the climate crisis—so clearly defies that linearity.

There is one thing that the climate alarm does do especially well, however, and one way that Thunberg's reference to the house on fire has been effective: it produces

> *an interruption*
>
> *a possibility*

an interruption. That interruption opens up a moment of possibility in which we might think differently about our response to climate change and the options before us. Climate change itself, after all, presents a problem for linear temporal models.[3] It is not only that cleaving too closely to progress and linearity aggravates the climate action that scientists demand, but also that climate

> *a problem of temporality*

change itself operates across multiple and discordant temporalities—what I called post-time in the previous chapter. Can Klein, Latour, and Malm's elaborations on climate warnings, as well as their meta-warning that the warnings are failing, help to address these points?

2. CLIMATE WARMING: A STATE OF EMERGENCY

In her introduction to *This Changes Everything: Capitalism vs. The Climate,* Klein, as her subtitle indicates, pits climate against capitalism. It is grievously "bad timing"

> *bad timing*

(16), she notes, that economic deregulation has been on the rise at precisely the moment we most need restriction and regulation. To cement the urgency of the climate situation she outlines, Klein borrows the metaphor of the house on fire from Pablo Solvo. "The various [climate] projections," she writes, "are the equivalent of every alarm in your house going off simultaneously. And then every alarm on your street going off as well, one by one" (15). Instead of acting in a manner commensurate to the alarm and "stopping the fire," however, "we are dousing it with gasoline" (14). The science has been established and virtually all climatologists, she notes, citing Lonnie G. Thompson, "are now convinced that global warming poses a clear and present danger to civilization" (15). Klein reads this clear and present danger as a warning: the climate projections *are the equivalent of* an alarm. The description *is* the performative. Nevertheless, she amplifies the warning by spelling it out: they are like a fire alarm. Not just one alarm: many alarms ringing down one's street. And we do nothing substantial to address the climate problem to which the alarm alerts us. "What is wrong with us?" Klein asks (14, 16).

What is wrong with us, she concludes, is that the power of corporations, and the neoliberal system that buoys them, impedes our ability to respond. We have not heeded the alarms because our political system makes it impossible to do so. Klein's meta-warning that the warnings are not working, then, embeds another warning: capitalism has defused the alarm. It has framed the climate warning in a way that makes the warning appear to be like Austin's bull in a painting. Klein returns to the rhetoric of a house on fire, the alarm, and the warning in *On Fire*. She concludes her introduction by citing Thunberg's call not only for a "new politics" and a "new economics" (53) in response to "the clanging fire alarm" (31), but also for "a whole new way of thinking" (53). But while Klein is heartened by the rise of the climate youth movement, as am I, she does not consider the new way of thinking in relation to time. In other words, deregulated capitalism may have arrived at an especially bad time, but it is also part and parcel of *bad timing* writ large that "a whole new way of thinking" about time might also address.

every alarm in your house going off simultaneously

every alarm on your street going off

a clear and present danger

what is wrong with us?

neoliberalism

capitalism

a whole new way of thinking

bad time
bad timing

In *Facing Gaia*, published just over two years after Klein's book, Latour, too, turns to the fire alarm to catalogue the failure of alarms and in doing so delivers his own meta-warning that the warnings are not working.[4] Like Klein, he repeats that the science is established. And yet "we" have not attended to the alarms:[5]

> How can we not feel rather ashamed that we have made a situation irreversible because we moved along like sleepwalkers when the alarms sounded? And yet we haven't lacked for warnings. The sirens have been blaring all along. . . . We can't say that we didn't know. . . . The alarms have sounded; they've been disconnected one after another. People have opened their eyes, they have seen, they have known, and they have forged straight ahead with their eyes shut tight! (9–10)[6]

eyes shut tight!

Latour later returns to this point with the specific metaphor of the house on fire:

> If someone tells you your house is on fire, whatever your indolence, your psychology, or your ancestry, you are going to rush outdoors, and the last thing you'll be inclined to do as you dash down the stairs is to stop on the landing to quibble about whether the firefighters who are setting up their big ladders are really firefighters and if they are 90 or 95 percent likely to get you out safely. If we were in a normal situation, the smallest warning about the state of the Earth and its feedback loops would have already mobilized us, just as any question of identity, security, or property would surely have done. (191)

the smallest warning

. . .

Latour implies that the warning is short-circuited by our skepticism with respect to the threat.[7] Warnings—the "smallest warnings" and, presumably, the biggest ones too—should generate action. But they do not. Acknowledging the climate warning's failure, Latour turns to speech act theory to gain a firmer understanding of the way warnings work. He notes, following Austin, that descriptions of climate change projections in the news are "intended as a *warning*" (emphasis his, 42). Michael Mann's hockey stick graph, indicating with the sharp rise of the base the rapid uptick in climate warming, consolidates this "link between description and warning" (43).[8] To see this image and to read the commentary that elaborates on it, Latour argues, is to be warned.[9]

The goal, it seems, on the part of Thunberg, Klein, and Latour is to borrow the clarity of action required for the house on fire and to apply it to our climate crisis. This is the force of Thunberg's *as if*. The projections *sound like* an alarm. But they are not like a fire alarm. They stipulate no action plan. Latour put it as follows: "What doubtless explains in part the old idea that [climate] description entails no prescription is that these [climate] warnings obviously do not spell out *in detail* what is to be done. They are merely ways of putting collective action under tension. Which is exactly what one asks of an alarm" (*Facing Gaia*, 48–49).[10] Arguably, they cannot indicate what needs to be done because we cannot collapse the complexities of climate temporalities to the clarity of the house on fire (or the bull in the field), a clarity that, Austin illustrates, comes with its own not inconsiderable demands. Further, we may recall that the warning is an illocutionary act whose action resides in its being issued. In this case, an alarm is rung to warn us. If the alarm is to be *effective*, however, a perlocutionary act—a consequence—must follow. We must not only be alarmed but we must also act in a manner commensurate to what the alarm dictates. And if no actions have been determined for the particular alarm we hear? We remain, as Latour aptly puts it, *under tension*. We are *alarmed* but we have no way to release that alarm into action.[11] In these cases, the alarm works not as a call to arms but only as a call to an anxious awareness of danger.

there's a bull! there's a hockey stick!

alarms are ways of putting collective action under tension

we are alarmed . . . as if

I tried to warn him but only succeeded in alarming him

alarms are alarming

The climate alarms have been alarming. This may be exactly what Latour wants from an alarm, but it is not what I want. Indeed, climate alarms have been effective in a different, and largely unintended, way that bolsters the neoliberal system that Klein identifies as the problem in the first place: they instill fear. And fear is a powerful vector through which neoliberalism operates. It enables, as we saw in the wake of 9/11, the passing of bills and waging of wars that would otherwise be met with greater resistance and dissent. It enables, as we saw then and continue to see now, the introduction of surveillance systems, the limiting of certain freedoms and the exercise of others, and it provides the rationale for a whole set of regulations, loosenings, and austerity claims that would have been unimaginable without the alarm. The alarms, in short, terrify.

We feel "under tension." In this context, the actions we most desire are those that will reassure us that we should not be afraid. This includes actions that contradict the alarm, throwing its very validity into question. If we are, as a society, investing in oil companies, for example, and

we could have acted bailing them out in the pandemic, then surely the climate crisis is not as bad as we think. If our friends are flying

we did nothing and living as they usually do and, importantly, *if we do too*, then surely the calls to alarm are not merited. For Latour is right that the projections are descriptions (constatives) that are also performatives (warnings). The problem is that because they do not dictate specific actions adequate to the climate predicament (and, as I will argue below, cannot), we seek solace elsewhere. It is a situation designed to make us all climate deniers for the welcome

what is wrong relief that such denial brings.[12] We want to believe that
with us? those projections do not mean what they seem to mean.

If we feel "under tension" in response to the climate projections that are also warnings, this tension is only heightened by the other common characteristic of climate warnings: the observation that they are not working, which is to say, they are not prompting action. In only a few short pages Latour catalogues the warning's failure to generate action and, in doing so, fuels its ability to alarm: we are "too late!" (*Facing Gaia*, 44), "too late" (45), "panic-stricken" (45), "stupefied" (45), "delaying" (45), waiting for "later" (46), and so on. With exclamation-laden astonishment Latour writes: we "haven't lacked for warnings" and yet "we hardly noticed a thing!" (9).

too late!
!!!

Like Klein and Latour, in *Fossil Capital* Andreas Malm laments our failure to act in response to the "permanent hailstorm of scientific warnings" (3). Like Klein and Latour, he also underscores that "everybody" is acutely aware of the climate problem we face and yet "the more knowledge there is of the consequences [of burning fossil fuels], the more fossil fuels are burnt" (3). Like Klein and Latour, his focus is on the meta-warning that warnings aren't working, a failure that he situates in relation to a capitalist ideology of progress. And like Klein and Latour, despite being keenly attuned to the complexity of climate temporalities, he still upholds the climate warning and, with it, the temporal model it supports.

everybody knows

Malm begins and ends *Fossil Capital* with the rhetoric of warning. Instead of the burning house metaphor, he adapts the speeding train from Benjamin's work.[13] The fossil fuel economy is a train hurtling toward a precipice, accelerating as it goes, the driver perhaps unaware or uncaring of the danger of the current course and, whatever the case, seemingly unable to stop the train (15). The train metaphor has some merits that the burning house does not. The train materializes our current situation. Malm does a nice job of tracing the fossil fuel economy back to industrial modernity and the rise of steam for which trains are a potent symbol.[14] The train tracks—or lines—also work as a convenient metaphor for linearity and progress.[15] The train, moreover, holds us captive as

the speeding train

rise of steam
rise of industry
rise of extraction

it takes passengers from station to station, further underscoring our powerlessness and the need for dramatic action if our current course is to be interrupted.

train tracks interrupted

Malm's book as a whole, however, is a call for collective and complex action that goes far beyond the train metaphor. Weaving together an argument that draws on ethics, representation, and politics, Malm seeks a more responsible, measured, and collective response to our current moment, a response that he threads through the complex temporalities of climate change: "Now," he writes, "is a singularly bad time for declaring the demise of time" (*Fossil Capital*, 10–11).[16] And his *The Progress of This Storm* returns to the centrality of time in its first sentence: "Is there any time left in the world?" But far from rethinking temporal models of the Global North—although very much committed to rethinking progress—Malm doubles down on their importance.

don't kill time!

is there time?
is there any time?
is there any time left?

Malm may use the metaphor of the speeding train to issue a warning about the need for climate action, then, but his larger work is a warning not to neglect the role of time and history in our response to climate change. Drawing on Gardiner's *A Perfect Moral Storm*, he notes that the ethical conundrum of climate change is at once temporal and historical: we are currently living, in a post-time sense, the consequences of what was done a hundred years ago, and what we do now is deferred for future generations. Malm is interested in the origins of climate change and the complexity that follows from its long timeline. In that context, the difference between the actions of our forebears and ourselves (say in the last thirty to forty years to present) is that *we know* what we are doing and they did not.[17] Still, it is difficult for many in the Global North to imagine these future people on whom our actions have such an impact. "The person who harms others by burning fossil fuels," Malm writes, paraphrasing Gardiner, "cannot even potentially encounter his victims, because they do not yet exist" (*Fossil Capital*, 8).[18]

those people do not exist

yet

This "messy mix-up of time scales" with its combination of "relics and arrows, loops and postponements that stretch from the deepest past to the most distant future" (Malm, *Fossil Capital*, 8) is difficult to represent; its messiness is something, moreover, to which the linearity of the warning is not adequate. Warnings are most effective when they are urgent, when the threat, like a house on fire or train going over a cliff, requires clear and immediate action.[19] But climate change complicates what both *action* and *now* means. The performative, to return to Austin, is a speech act that happens as it is spoken: to speak it is to perform. Climate warnings warn (they are illocutionary acts) but the perlocutionary effect is left hanging. The fact that the warnings are often issued in relation to longer time spans than the present moment, moreover, also enables their infinite deferral. These multiple time scales—which are not unique to climate change but only more pronounced and consequential in its context—do not guarantee predictable outcomes (evading the bull, putting out the fire).

And yet Malm steps back from the full implications of his argument and submits to the same model that has created the problem in the first place and is, I am arguing, inadequate to its redress. Malm returns to the metaphor of the speeding train as climate warning in the last chapter of *Fossil Capital*, entitled "In the State of Emergency." Can we, he asks, drawing on Walter Benjamin's work, pull the emergency brake "to derail the ultimate disaster of the present"? (394). The "prospects are dismal" for responding to the obvious climate crisis but, he allows, we might yet "spring into action" (394). The chapter begins with Benjamin's 1928 comment that society's tendency to cling "to normal (but already long lost) life is so fierce as to frustrate the truly human use of intellect and foresight, even in the face of drastic danger" (cited on 392). We might say, inverting Latour, that the "smallest [and largest] warnings" are likely to be disregarded when they

messy, messy

warnings warn

they do things with words

what?

even in the face of drastic danger

threaten "normal life." We have seen this point realized again in the pandemic. Just over ten years later, Benjamin reflects on the atmosphere of anxiety and crisis that pervades Europe. The "'state of emergency' in which we live is not the exception but the rule," Benjamin claims (cited on 392). Malm extends this sense of pervasive crisis back to the beginning of the Industrial Revolution and forward to our current climate emergency (392–93). The alarms have been rung, and the result is a constant state of emergency. Another word for this state is *panic*. And, indeed, panic is precisely what Malm wants to cultivate in his reader. "Dare to feel the panic," he writes (*Progress of This Storm*, 226), anticipating Thunberg's much-repeated injunction, "I want you to panic."

panic
panic

But inflammatory, panic-inspiring warnings may be ill advised. When we are warned, we are alarmed. Is it helpful to warn us to act as we would if we were in a house on fire? To where should we rush? To whom should we turn? Most importantly, how does one wrestle with the temporal complexities of a crisis that is immediate in multiple places but distant in others, that requires immediate action everywhere even when the crisis itself remains largely invisible and will only emerge, full force, at some unknown future point? Nobody seems to know. The meta-warnings with their dismay and admonishments—Klein's "What is wrong with us?," Latour's stirring up of shame—may serve only to heighten our alarm. This response not only fails to inspire effective action on climate change but also arguably facilitates and underwrites the very neoliberal framework that frustrates such action.

we are alarmed

And yet: we have to know that a crisis exists. The metaphor of the house on fire contributes to that work. It translates the science into terms that are readily legible to the public. And if to know is to warn, and to warn is to alarm and frighten, then isn't a sense of pervasive crisis and alarm only an inevitable condition of the necessary mode through which climate change will be addressed (if clumsily and too slowly)? Or is there another possibility? Klein, Latour, and Malm each note, without fully addressing, the interleaved, multiple temporalities specific to our comprehension of climate change in combination with the "bad timing" of the neoliberal political configuration of the Global North. The warnings and meta-warnings they issue, however, have a compelling antecedent in Benjamin's work, as Malm notes. Benjamin's rethinking of temporality in tandem with his materialist history draws out a point that all three of these anatomies of alarm neglect: the privileging of interruption as a critical mode to address social and political crisis. In doing so he divides the potential work of the alarm between a call to action, an anxious awareness of danger (being "under tension," panicking, the ongoing emergency), *and* a real state of emergency that interrupts all of these definitions by interrupting the structure through which they are understood and instead locates action in forms of response that reconfigure existing temporal models in new ways.[20]

a real state of emergency

3. "QUITE OTHERWISE": A *REAL* STATE OF EMERGENCY

The power of Benjamin's work is to open the performative contradiction that at once warns and disarms the warning. In the interval between the warning and the recognition that warnings fall on deaf ears and produce only a state of emergency, there is the possibility of a different response. Importantly, this gap is rendered materially and formally and follows from Benjamin's appreciation of how literature works; indeed, his own philosophical and political practice, as I have noted, is literary in this way. While Malm, like many other critics, has found Benjamin's metaphor of the speeding train resonant for climate change thinking, he does not note this passage's context. It follows almost immediately on the heels of Benjamin's claim that a classless society is not the goal of progress but

Alarm:
- call to action
- anxious awareness of danger
- a real state of emergency that interrupts

its interruption (402). This interruption is materialized in the metaphor of the train that follows:

> Marx says that revolutions are the locomotive of world history. But perhaps it is quite otherwise. Perhaps revolutions are an attempt by the passengers on this train—namely, the human race—to activate the emergency brake. (*Selected Writings*, 4:402)

revolutions
get off the train
activate the emergency brake

in time?

Malm reads the implicit warning in this statement and asks: Will we act in time? Will we get off the train? Michael Löwy, turning to Benjamin's own invocation of the fire alarm, also asks what we can learn from Benjamin's recourse to the alarm; he suggests that his "whole work" may be interpreted "as a kind of 'fire alarm' to his contemporaries, a warning bell attempting to draw attention to the imminent dangers threatening them" (*Fire Alarm*, 16). But I wonder if Benjamin's warning can be understood in a different way. Instead of pulling a lever and sounding an alarm that is a call for action (say, evade the bull, leave the house, put out the fire, or get off the train), it sought to put the pressure less on the action that followed the alarm and more on the call for *interruption* itself. We can trace Benjamin's thinking about this possibility from his early work through his last writings.

we interrupt
we are interrupted

one-way street?

In 1928, Benjamin published a short, idiosyncratic text entitled *One-Way Street*. Fascinatingly various, this work defies academic form to embrace instead the heterogenous, formally experimental, kaleidoscopic style that would become a hallmark of his writing.[21] Informed by surrealism, Baudelaire's prose poems, and what Andreas Huyssen calls the "metropolitan miniature," *One-Way Street* is composed of sixty short, boldly titled prose pieces. Many of these pieces had their origins in the new feuilleton structure of newspapers and magazines in which the bottom third of the newspaper's page was dedicated to cultural and political commentary, serialized commentary, and sundry observations (Jennings, 258); their placement "below the line" (the translation of *unter dem Strich*), Michael Jennings suggests, would likely have appealed to Benjamin.²²

below the line

About two-thirds of the way through the text, Benjamin includes a short piece entitled "Fire Alarm." While most critics, encouraged by his book's title, focus on the spatial topography of *One-Way Street*, this particular entry relates to time and Benjamin's views on temporality. Both Rebecca Comay and John McCole read "Fire Alarm" in the context of the state of emergency captured by the surrealist alarm clock "that in each minute rings for sixty seconds" (Benjamin, *Selected Writings* II:56). Seemingly interested in class war and its interpretations and possible outcomes, the section closes as follows:

> And if the abolition of the bourgeoisie is not completed by an almost calculable moment in economic and technical development (a moment signalled by inflation and poison-gas warfare), all is lost. Before the spark reaches the dynamite, the lighted fuse must be cut. The interventions, dangers, and tempi of politicians are technical—not chivalrous. (84)

before the spark reaches the dynamite

The structure of "if . . . then" adopts the structure of warning but it deliberately holds back from suggesting a time for action. The action, however, is specified: "the lighted fuse must be cut" (84), the line must be cut. What is the titular fire alarm? It seems at once to alert us to the "lighted fuse" and the explosion to come. As Comay notes, for Benjamin the "continuum—continuation as such—is the catastrophe" (262).²³ That is, the very model on which the fire alarm relies is the catastrophe. Benjamin may also be trying to distinguish the dangers, interventions, and tempi on which the bourgeoisie rely from the very different interventions, dangers, and tempi of radicals. In the former, the line is maintained even if there are

*the lighted fuse
the fire alarm
the alarm clock
the explosion
to come*

below the line occasional breaks; in the latter, by contrast, we are "below the line," out of line, in a different zone all together.

Consider Benjamin's radio broadcast on the Tay Bridge train disaster delivered in 1932 with which I began this book. On an evening in late December 1879, a train carrying seventy-seven passengers from Edinburgh to Dundee plunged into the River Tay. The new and massive bridge, constructed from iron and tested by engineers, had given way as the train moved through a winter storm and high winds. There were no witnesses. This drama was much recorded in the press of the period, and it is not surprising that Benjamin would return to it to think about new technologies—the rise of steam, the use of iron, Stephenson's invention of the locomotive—and industrial modernity. Benjamin's response is a mix of wonder and dismay prompted at once by the accident and "a gaping hole in the line" (*Selected Writings*, 2:567).

a gaping hole

Several critics have illustrated how representations of nineteenth-century train accidents serve as forms of consolation and management, on the one hand, and petitions for agitation and policy, on the other. I want to stay, however, as Benjamin does, with "the gaping hole."[24] For accidents, like alarms, also interrupt. Unlike alarms, however, the challenge is to stay with the interruption, hold it, and work it. The moment of interruption presents different possibilities, however fleeting and unrealized they might be. Further, while Benjamin's radio broadcast does not lend itself to registering the accident in its form, his print warnings—from the fire alarm in *One-Way Street* through the train and the angel of history—are notably issued in works that also dramatically rethink traditional forms.

a poem, a text, an image = angel = warning = the failure of warning = poem, image, text =

I considered the reference to the angel of history in "Interruption" in relation to the absent reproduction of Klee's painting and to the fact that Benjamin's text, while presenting itself as a kind of ekphrastic description of Klee's image, bears very little resemblance to that image.

But can this image, as described by Benjamin, also be read as a warning? The angel stands transfixed before calamity. He appears to be capable neither of learning from the history at which he stares nor directing his action into the future toward which he is propelled. "The angel would like to stay," Benjamin writes, "awaken the dead, and make whole what has been smashed" ("Theses," 257).[25] But he cannot. He is being thrown into an unseen future with such violence that he cannot even close his wings. He wants to warn others (the fire alarm), wake them up (the alarm clock), and fix the damage (the train). But he cannot issue these warnings. Indeed, he warns about the failure of warning: the angel cannot do his work, he cannot awaken the dead, he cannot tell us what he sees. It is too late.[26] The angel's traditional role as messenger no longer obtains. This is not a description of a disaster to come (say, the hockey stick graphs) but rather a description of the failure of the ability to warn of that disaster.

awaken the dead
(the alarm clock)

the smashed
the speeding train
it is too late

But consider that the warning against warnings *is* issued. The performative contradiction holds the possibility and impossibility of warning together in the same moment. Austin and Latour, as noted above, are both attuned to the ways in which the constative and performative are unstable. Reading Austin, I am struck by the care he takes in moving through his examples and his willingness to pause and adjust his categories when they do not meet his previous claims. *How to Do Things with Words* is an essay that adopts a conversational mode to build an argument about language, an argument that in the last third of the book Austin just as patiently takes apart.

I warn you . . .

how do to things with words

"Theses" is a very different sort of work. The theses are not theses, per se, but short, intense stabs at the preoccupying ideas of Benjamin's intellectual circle. They do not build on each other and they do not produce any sort of theory (to be taken apart or not) in their conclusion. If anything, they are implicitly combative, colliding with each other and sending sparks across the page that Austin's more measured work avoids. And yet I have wanted to bring them together because Benjamin's essay, and indeed his work as a whole, is so palpably performative. It is not only that sparks fly, but also that his work is often combustive if not explosive. It makes things happen even as it, as in the dialectical image of the angel of history, illustrates how the moment for action has passed. In short, it takes us out of the continuous emergency and helps us to gain an appreciation of the *real* emergency as it elaborates a technics of interruption through a savvy manipulation of form. Consider again the angel of history: to address it, Benjamin offers Scholem's poem, an ekphrastic description of Klee's image, and his own exposition of it. They collide and create gaps. The thesis form and the temporality it asserts is *about* those dynamic, relational gaps. Those *stops*. Those gaping holes.

Those gaping holes. . . . As I write, the climate crisis is only intensifying, and the need for immediate action is only becoming more urgent. What the moment seems to call for is not hesitation or pausing, but a shoring up of knowledge as a catalyst for action. I began this book by noting that the climate change idea is still very much in the process of formation. The dissemination of climate science and the hard planetary boundaries it brings into view, coupled with the many different arenas—cultural, journalistic, visual, activist, and so on—in which the climate crisis has been represented, have been compelling, and often moving. But they have not been action-propelling in the way that counts: the reduction of carbon emissions. This impasse invites a multiplication of approaches and a recognition that the very temporal frames for thinking to which the Global North is often indebted, and on which capitalism relies, may, indeed, compound the difficulty of generating climate action. I am now some distance, however, from one of the most prominent ways in which Benjamin's

---

make sparks fly

the real emergency

gaps
stops
gaping holes

is there time to pause?

hard planetary boundaries

x x x x x x x x +

distinction between the state of emergency and the *real* state of emergency is taken up in critical studies. This interpretation emphasizes not emergency but exception and argues that a state of exception suspends rights, asserts power, and subdues populaces on whom that power is exercised. This is a problem to which Agamben, following Benjamin, rightly points.[27] The challenge, then, is to rethink how emergency *can* work, how a *real* state of emergency can be realized.[28]

In this context, I wonder if the question is not, Is there time to pause? but rather, Is there time *not* to pause? Counterintuitively perhaps, to pause *is* to act. This action is not oriented toward some distant end: it happens now, in this moment. That said, the pause does not offer a plan. It creates a catalytic space where plans may be made and unmade otherwise, where time may be made and unmade otherwise. I understand the real state of emergency—a state that Benjamin resists demarcating in any straightforward way—as refusing the continuous alarm and at once performing and responding to a different sense of time. The "real emergency," then, lays the groundwork for that "something new" calibrated to different temporalities. The four experiments in the next section flow from my exploration of interruption, post-time, and the real emergency in these Beginnings. They are very provisional examples of the "otherwise"s to which I gesture in the Preface and throughout (following Benjamin and others). I try this out as an academic but, in general, this is a call for all of us, in the context of climate precarity, to put pressure on, and/or reinvent, the idea of time in which we work, think, and act.

Klein, Latour, and Malm all issue the meta-warning that climate warnings are not working. But, while sympathetic to the temporal complexity of climate change, they do not pursue Benjamin's additional emphasis on interruption and its promotion of a different temporality that both representation and warning, traditionally

is there time to pause?

*is there time **not** to pause?*

plans made and unmade

time made and unmade

a catalytic space, an x x x of possibility

otherwising other ways

understood, cannot meet. Interruption offers a different response to our current situation, one that attends to form and produces not the continuous emergency of the fire alarm but rather the "real emergency" that makes action possible. Like Malm, Löwy links the metaphor of the speeding train to climate emergency in his reading of Benjamin: "We are witnessing in the early twenty-first century the ever faster 'progress' of the train of industrialist/capitalist civilization into the abyss, an abyss called ecological disaster, the most dramatic expression of which is global warming" ("Revolution," 3). At the close of *Fire Alarm*, however, he elaborates on this point in a register that approaches what I am suggesting here: "Benjamin's fire alarm retains its currency to a striking extent: catastrophe is possible—if not indeed probable—

unless . . . unless . . . Though formulated in the style of the Biblical prophets, Benjamin's pessimistic predictions are conditional: there is a danger of this happening, *if* . . ." (Löwy,

if . . . 111). Lowy's ellipses invite reflection without naming an action or issuing a warning that could approach the terrain of the "too late." These ellipses, these pauses or interruptions, signal the "open conception of history" (113–14) for which Löwy, following Benjamin, is advocating, a history that invests not in teleology but rather in open, co-written, and multiple possibilities.

4. CONCLUSION: THIS IS NOT A WARNING!

a teenager speaks On January 21, 2020, just before the global pandemic was announced, Thunberg returned to the World Economic Forum to speak again on climate change. She recalled her speech from almost a year earlier and her petition to world economic leaders to act as if the world were on fire. Her frustration was clear. Nothing has been done, she noted, in the intervening year. "Our house is still on fire. Your inaction is fueling the fire by the hour." Behind Thunberg, the words "Averting a Climate Apocalypse" were displayed on a large screen over an image of a girl, carrying a bundle on her head, walking across a parched landscape with her back to the audience. Beside that image was the now iconic silhouette of the kangaroo hopping before raging fires in Australia. Thunberg implored her audience of economic leaders to act: "We want this done *now*." They watched and listened.

a girl walks away

a kangaroo hops

People criticized her call to panic, she remarked. But there was no need to do so. No one acted. "It didn't matter," she said. The panic had no impact. I have sought to convey in this chapter, by contrast, that the rhetoric of warning and alarm does matter but not in the way the way that Thunberg intended. It creates an "anxious awareness of danger" in combination with no clear agenda for action because our temporal models are inadequate to climate temporalities. The alarm produces a fear that seeks relief in structures of denial and avoidance. Thunberg's message was not powerful enough to disrupt the solace many people find in "normal life," which—for those in the Global North—is also the neoliberal economy.[29] In words that now carry an added resonance in the midst of the pandemic, she noted that climate action "seemed so bad for the economy" that as a world body we did not even try to address climate change. Indeed, the alarm incited by the climate warning perhaps unwittingly led to a *bolstering* of these very economic structures that require transformation because, under neoliberalism, they offer a sense of security that is only sought more desperately when it seems under attack.

still

still

normal life

Thunberg returned to the metaphor of the house on fire, with a note of resignation, just over a month later, in a speech delivered before the European Union on March 4: "The house is on fire, but people have gone back into the house, watched a movie and gone to bed. The fire brigade has not even been called to put out the fire." While many were sleeping, however, Thunberg's words sparked a worldwide movement, brought more people to the streets in climate protest than ever before, politicized youth, and may yet make a dent in those economic bodies that currently have such a hold over fossil fuel emissions.[30] But as of this writing, that coordinated global response to the climate crisis has not come close to being realized. Part of the point I am making—in this chapter and throughout this book—is that the metaphors we use, the stories we tell, the rhetorical strategies we adopt, shape how we imagine and respond to climate change. Thunberg began with a climate warning: "Our house is on fire." When she returned to this warning a year later, it had become a meta-warning like those we have seen in Klein, Latour, and Malm: "Our house is still on fire." The "still" in the second warning is the meta-warning that warnings have failed. But in the gap between those two speeches, a "quite otherwise" also opened.

Consider again the stage from which Thunberg made her second address to the World Economic Forum. The audience watched not only Thunberg but also the words and images behind her. A young girl, with her back to the audience, walks away. A kangaroo hops. If we recall Benjamin's angel of history, we could perhaps consider Thunberg as the messenger who is still empowered to warn us, delivering her message on "averting the climate catastrophe." At the same time, however, another girl walks away. She turns her back to the audience, at once preserving and cancelling Thunberg's words. She attends to her daily task. Thunberg talks to world leaders. Thunberg can warn, the kangaroo can manifest the reality of that warning. But perhaps the gap between the girl walking away and the girl warning the world is the gap—the gaping hole—that a theory of interruption animates, the ellipses, the . . . , the pause of possibility that it is now our task to hold and work.

. . .

PART II

On Time: Four Experiments

On Time: First Experiment

Layering

This chapter is composed of six bands that imperfectly reproduce the geological stratification offering insight into deep time. Each band can be read on its own or in combination with the others. Either way, I hope that readers will consider the effect of the layered presentation on the page. The point of many cultural forms—and especially academic forms—is to be invisible, to not get in the way, so that the ideas and images can be communicated smoothly without interference or static. The success of these forms, however, means that we also often forget them. Visual and literary artists have long explored the legibility of form in their works, drawing our attention to paint and materials, the composition of images, frames, words on the page as images, words-within-words, white spaces, and so on. Academic writing, however, has been slower to introduce an attention to form into its practice.[1] In *Anthropocene Reading*, Menely and Taylor ask "how the Anthropocene might require us to read differently" (12). In this chapter, I ask how it might also invite us to *write* differently and to explore new forms for imagining the climate change idea and the moment we occupy in relation to it.

Stones are a good place to start. They have shape, substance, weight. They are form forward. To read them also requires a certain set of skills; most of us, for example, do not pick up a stone and ask where it falls in stratigraphical surveys and deep time ledgers. And to use stones and rocks and the layered bands that geologists have developed to think about writing and words can seem even more counterintuitive. But here I follow the lead of Marcia Bjornerud and other geologists and literary critics, not least among them the critics collected together by Menely and Taylor in *Anthropocene Reading*. Bjornerud, for example, brings both geology (her field) and literary studies together; literary studies, she writes, involves "close reading" as does geology. Indeed, she writes, geology is "nothing less than the etymology of the world" (21–22). To illuminate this etymology, the "palimpsest" helps her to think about new ways to "organize time" (22). The palimpsest does not preclude a narrative history but rather organizes it otherwise. Instead of striated rock layers, for example, there is *at once* a series of lines or layers and an overlapping in which those lines or layers occupy exactly the same space and time. Robert Pogue Harrison's etymology of *logos*—the word—is useful here; it harkens back, he writes, to "*leg*, which meant 'to gather, to collect, to bind together'" (72). A gathering, collecting, and binding together conveys a different sense of meaning than a line that struggles to capture simultaneity and implicitly promotes a sense of progression.

The bands I develop in this chapter are also a gathering, collecting, or binding together. They at once speak to one another and speak within and across time. In the two middle bands—bands three and four—I revisit material I read as a graduate student just over thirty years ago. Somewhere in the process of writing, I realized that my academic career spans the period when the climate crisis increasingly came into view in the Global North. And so when I return to these works, they are something new and different to me now, read through the prism of climate change.

Ecocide may not have informed my early readings, but suicide and accidental death traverse, in different ways, all six bands. As if to write on the climate crisis and ecocide is also to write on its gut-clenching avatars, suicide and death. Like deep time and climate change, death brings representation short. Which does not mean that attempts at representation are not made. There are two books on the anthropocene entitled *Learning to Die* and many others that share this theme.[2] But there are also writers who use this moment of climate peril to think about death and deep time in new ways. Bjornerud, for example, suggests that the tiny sliver of time occupied by humans on the earth's timescale—our "second before midnight"—is a spur to thinking otherwise about both time and the human.[3] Thinking about the minuteness of the human, Bjornerud writes, fails to account for the magnitude of human impacts on earth systems, fails to account for our deep and indelible ties with times before the human, and, with its metaphor of midnight, suggests that there is no future (6–7). Macfarlane similarly suggests that to see human existence as a "blink of a geological eye" should provoke "us to action not apathy." "For to think in deep time," he writes, "can be a means not of escaping our troubled present, but rather of re-imagining it ... with older, slower stories of making and unmaking" (*Underland*, 15)[4]

In the nineteenth century, geologists and writers struggled with those stories of making and unmaking. Martin Rudwick and Adelene Buckland, among others, have illustrated how James Hutton, Charles Lyell, and Charles Darwin, for example, created new visual and literary forms in an effort to capture what was beyond human comprehension. As Buckland puts it, geology "was written into existence in the nineteenth century as much as it was found" (4).[5] This metaphorical traffic also moves in reverse, with critics like Benjamin using the rock record as a metaphor for epic forms and pitting those forms against new print forms. As Menely and Taylor illustrate, "modernity's accelerated tempo" was, for Benjamin, at odds with "the dilatory rhythms and extended durations" of geological time (1).[6] Benjamin's comments then become a point of departure for thinking about "the shaping power of narrative to organize time" (1). While the rocks had a story to tell, it was the narratives and images and rhetorical devices that scientists chose that made that story legible to mid-nineteenth-century readers and that continues to inform our understanding today.

I am inspired by how some critics, seeking to find new forms for this story, have pushed the boundaries of the academic essay. In a formally interlaced story and argument, Jeffrey Jerome Cohen, for example, like Bjornerud, foregrounds the palimpsest to convey a "new relation to time" (162). The palimpsest captures a "gyred" (35), "cork-screwed" (29), "vorticular topology of reading" (26) that does not seek "one text layered upon another," Cohen writes, "but time convoluted into a whorl" (35). This idea chimes with the Convolutes of Benjamin's *Arcades Project*, which, among other things, is also an attempt to find a form for a new relation to time. Robert Smithson's "nonsite" art installations are also helpful to consider here. They take up the question of time, refusing "timelessness" and advocating something more like the "timefulness" that Bjornerud develops in relation to geological studies. Smithson's work, Lewis Hyde writes, opens "an inquiry into how we think of time" (276). His engagements with stone, earth, and quarry suggest to me that representations of climate change can be ever-renewed makings that do not settle to in modes of completion. And yet, at some point, they can *settle enough* to be intelligible while still continuously open to being unsettled again.

This chapter also invites a reflection on the material modalities of writing and reading. The poet Mary Ruefle writes by longhand because she wants a material connection between her body, the pencil, and the page—what she calls "the whole wrist-moving action" (cited in Tricker)—to *inform* her poems. Most academics, and probably most poets, however, use a computer to compose their works. I wrote this book, for example, on my computer. But this chapter posed challenges for my Word program; Word did not let me write in the layered bands I had envisioned (indeed, to navigate the problem my nephew, Noam Flear, wrote me a computer program that circumvented it). Having my attention drawn so forcefully to the ways in which the tools of writing dictate *how* I write also served as a reminder to think about how the tools of reading dictate how we read. For this chapter, in particular, the material modality of reading—printed page, bound book, e-device, audio device, and so on—will make a difference.[7] This point is often forgotten when we use familiar forms, but even the familiar forms are *always there*, of course, and no less so than when they disappear from conscious awareness. This is a point worth bearing in mind not only as we read but also as we draw on, as we always do, that most invisible of material forms, energy systems that create carbon emissions.

In *Timefulness* Bjornerud catalogues several art works that seek to expand temporal models beyond human frames. "It may be tempting to dismiss these projects as gimmicks or follies," she writes, "but their purpose is to reframe the way we think about ourselves in time" (169). This chapter, too, may seem like a gimmick or a folly, but my purpose is to reframe the frame itself: to ask readers to see and experience it. My hope is that the process will, as Bjornerud desires, also generate different ways in which time may be experienced *and* different ways that time and climate change may be understood. My goal, here and elsewhere, is to present climate change as an idea that, like all scientific ideas, can be represented in a range of ways, each of which has an impact on how it will be understood and addressed; to underscore that frameworks for understanding ideas like climate change are, by definition, never absent or optional; and to suggest ways in which our words and forms can use interruption, think and *work* interruption, to make visible the frames that we are in the habit of overlooking and to attend to questions of framing and interruption in the service of new frames and new possibilities. In short, I want to manipulate academic forms—that, like all forms, are part of our analysis whether we foreground them or not—in an effort to incite new, more inventive, ways of making and unmaking time and the climate change idea.

. . .

Explanatory note on the formatting: (1) With the exception of this short introduction, this chapter embeds its references within the text and includes no footnotes.[8] (2) Paragraph breaks are indicated by degree symbols. (And [3] one note on formatting that is not here: I had initially hoped to have all the *and*s in this chapter printed in blue—the *and*s themselves and the *and*s in s*and*, b*and*, h*and*, and so on. This idea proved too difficult to execute but, as you read, see, if you can, these *and*s and the possibilities they invite.)

BAND TITLES

Band One: Grass, Epigraphs
Band Two: Bad Weather, Boats
Band Three: Origins, Words
Band Four: . . . .
Band Five: Origin, Rocks, or How to Read
Band Six: Erosions, Sand

Band One. *Grass, Epigraphs.* "The war and the constellation that brought

....

Band Two. *Bad Weather, Boats.* In mid-July 1822 two partially decomposed bodies were found washed ashore on the beaches of Italy's Gulf of Spezia. One of the bodies had been hastily buried in the sand, and in the coat pocket of the other body was a slim volume of Keats's poetry. A third body was found on a nearby shore a few weeks later. These were the bodies

....

Band Three. *Origins, Words.* My idea to present this chapter as a series of sedimented, layered bands derives from Jacques Derrida's "Living On/Border Lines." This essay has stayed with me since my graduate student days. I remembered that Derrida had written the essay in two running bands, the upper band occupying roughly two-thirds of the page and the band below a third. I remembered that the essay addressed Blanchot. But returning to it for this chapter was a surprise. For not only did it experiment with form (what I remembered most), but it also addressed life, death, afterlife, and

....

***Band Four* . . . . .** Punctuation tells us how to read. Stop here, pause for a breath there, lift your voice here, feel the drama there. The period or dot signals the end of a sentence. Stop here. But the period, period, period or dot, dot, dot signals the opening up of a sentence. It places a small internal tear in it, a rip. It gestures to what cannot be said, what is left out, what will be silently passed over, as well as to an emphatic temporal pause (and

....

Band Five. *Origins, Rocks; or, How to Read.* Initially, I envisioned this band to follow the epigraphs on grass. One of the epigraphs to Lyell's *Principles of Geology* is from Linnæus: "The stony rocks are not primeval, but the daughters of Time." I was interested in how rocks embodied time, how

....

Band Six. *Erosions, Sand.* "The sands are numbered that make up my life" (Shakespeare, *Henry VI*, part 3, i.4). "To watch grain after entrained it about led me to take down a few thoughts which I can say that I have

....

of Percy Shelley and his sailing companions, Edward Williams (with whom Percy was renting Casa Magni with his wife, Mary, and Williams's wife, Jane) and Charles Vivian (a young sailor from the area). Shortly after setting out on their seven-hour return journey from Lerici to Casa Magni, their boat had capsized in a sudden storm. This tragic boating accident is

....

temporality in ways that resonated with ideas I am pursuing in this book. It was a surprise, too, to see the many handwritten comments I had made in the margins and between the lines in that period of reading thirty years earlier. But it was mostly a surprise for its engagement with Shelley's "Triumph of Life." If I remembered anything about the essay, I should have remembered that. ° Derrida opens his essay as follows: "But who's talking about living? / In other words on living? / This time, 'in other words' does not put the same thing into the same words, does not clarify

....

then . . .) and an interruption in one's train of thought. But like most punctuation, we read and do not read it, we take its direction and carry on. One of the many things I love about Virginia Woolf's writing is its attention to the mediation of punctuation. As Samuel Weber notes, drawing on Benjamin, punctuation, when we see it, approximates pure language; it is simply there (78). When I came across this sentence in *A Room of One's*

....

their stratifications were read, how one could hold a stone in one's hand that defied one's mind's capacity to encompass its time. If I could not quite inhabit deep time, I could sense how stones were like fallen stars. In relation to time, rocks make mediation—the medium—inescapable: time has a

....

sand grain roll, bounce, and be carried aloft. Long-avoided questions emerged as the current nudged me downstream" (Savoy, 16–17). "But kept with me, indeed kept from myself, for nigh on twenty years. [. . .]⁹

...

well documented. °   Percy's friend, Edward Dowden, in an account entitled "Last Days" in his *Life of Percy Bysshe Shelley*, describes the storms and winds in the weeks preceding the accident and then the dry weather, the oppressive heat. "Through storm and calm the days went by [. . .]. Now a gale encircled the bay with whirling foam, and the waves broke

...

an ambiguous expression, does not function like an 'i.e.' It amasses the powers of indecision and adds to the foregoing utterance its capacity for skidding" (75). In some ways, these passages are about commentary and its proliferation—"'in other words,' and so on and so forth" as Derrida notes a few lines down (75)—but in other ways these passages are about time, "this time." Derrida's reading is amplified in the band below the text which begins with the title and then the date. And in still other ways, the essay is about "indecision" and "skidding." It holds these meanings and others

...

Own I stopped and read it again: "For truth . . . these dots mark the spot where, in search of truth, I missed the turning up to Fernham" (15). Woolf brings the ellipsis into view and asks us to read it. Ever since I encountered that sentence, now many years ago, I have wanted to write an essay entitled only ". . .". This band is not quite that essay, but it develops Woolf's reflections on ellipses to address one of the conundrums of interruption,

...

weight, an appearance, a color, a size. I could hold it in my hand. But the idea behind this chapter was that, in the case of geology, time is also striated. Rocks are read in rows. (These linear bands remind me that almost all of our reading in the Global North is in bands, although the printed text is

...

at length these forests and grassy plains were consumed, being suddenly blown into the air, and their ashes scattered to the winds" (Lyell, *Principles*

Even today, I am handing them to you more as a bouquet of whispering

....

upon the beach with a sound like that of booming artillery; now the waters covered the nautili [. . .] which told of a lull after wild weather" (552). One night while Percy was standing on the terrace of Casa Magni with Edward, watching the sea, he stepped back in sudden horror. He had seen, he told his friend, the small head of Allegra (the recently deceased

....

together and does not resolve (settle or decide) them. It lets the meanings skid.   °   The second paragraph begins: "In other words on living? This time it sounds to you more like a quotation" (75). Not only is the earlier question repeated—quoted—but so, too, is "this time." "This time" becomes a new time, a second time. The problem of time is reinforced when Derrida refers to the passage as "out of joint" and in doing so extends the context outward, backward, to Shakespeare's time and Hamlet's time of indecision, a time of skidding. He also extends it to us, his readers in the

....

the way that it lists in two directions: as a disabling distraction that prevents thinking and as a radiant disruption that opens a space for new thinking.   °   "For truth . . . these dots mark the spot where, in search of truth, I missed the turning up to Fernham" (15). The dots—the unrepresentable in language—mark the spot where one searches for truth and misses it. It is not here that one *finds* it, but rather it is the place

....

not usually considered in this way.) Each row or band of rock, moreover, is a compression of time and invites a kind of play with deep time. It is discovered under our feet, deep in the earth.   °   As I was writing this chapter I chanced upon a sign, almost buried in snow, outside a series of

....

[e-book], 374). "And fragrant zephyrs there from spicy isles / Ruffle the placid ocean-deep, that rolls / Its broad, bright surges to the sloping sand"

grasses, gathered on reflective walks, than a collection of theses" (The

....

five-year-old daughter of their friend, Claire) bobbing in the waves and smiling. °   Dowden's account of Percy's "last days" is informed, as this phrase underscores, by his knowledge of Percy's coming death. He is alert to premonitions: the child's face bobbing in the waves, Mary's plea to Percy to stay home, the many people who saw Percy's "ghost" in the weeks

....

2020s, and our time, which is also "out of joint," a time in which *living* is a word that, for many, now has a small inflection, a lift, a how long will I live?   °   Taken together, the two bands remind me of mediation, but the band below goes a step further to put a focus there: it is written in a "telegraphic style," it invokes "telegraphics and telephonics," it can be read as "a telegram of a film for developing" (77). My graduate student self underlined these passages. We are always reading between the lines of what we have read before as well as what we have not read. And yet

....

of its possibility. And that possibility—the . . .—is a gap, a departure, a disruption, an interval, an intimation of something to come that does not arrive. And yet it is also there: ". . .". The dots are the place where the narrator, Mary, absorbed in her thoughts of truth, goes off the path, gets lost. The dots are also the place where we, as readers, are interrupted in our reading, are thrown and momentarily disoriented. I'm sure when I first

....

sprawling, midsized, brown brick buildings that were the offices of Geological Survey of Ontario. It was a bitterly cold day, and the streets were deserted. But, even so, it seemed unlikely that these generic buildings, resembling abandoned hospital buildings or army barracks, would ever be inviting

....

(Shelley, "Queen Mab"). "To sediment is to be alive in a timely fashion, where time isn't abstract, mathematical, monetary, or linear but always

"Theses" letter of introduction, from Benjamin to Gretel Adorno, May

....

before he died (as did Percy—he was walking on the terrace and stopped short when he saw another figure he recognized as himself). All of these details are recorded through the prism of this later knowledge. Before the accident, these were the details of daily lives; after the accident, they became details that portend the trouble to come.   °   To allow the

....

these unremembered echoes from my past reading about the afterlife and suspension produce an uncanny sense of suspension themselves, as if I am on a suspension bridge between two time periods and feel the sway of the bridge becoming my body and sense the gorge with its rushing water beneath me.   °   In the band below, Derrida muses, "I shall perhaps endeavour to create an effect of superimposition, of superimprinting one text on the other" (83). "In other words" imprinted on "In other words," although he does not give this example. I think about the geological term

....

read this sentence, unfamiliar with Woolf's stylistic idiosyncrasies, I wondered, what dots? And then realized, *those* dots, the dots on the page. Woolf brings us abruptly back to the materiality of language here, reminds us that the search for truth is conveyed through language, and reminds us that as close as we can come to truth is the place were language breaks, pauses, makes a space between the words. When I returned to Woolf's

....

to passersby. And then I saw the sign, in standard government format, that read like poetry: "Cobourg beds, / Ottawa formation / Barneveld stage, / Ordovician / These / sedimentary / rocks are / about / 440 million / years old." It recalled Smithson's "nonsites" that ask us to attend to "temporal

....

embodied, coalescing, 'eddying' into a story about itself. [. . .] To sediment might mean to sleep wakefully" (Lemenager, "Sediment," 179). "[Poetry]

1940). "Hold out your hands and let me lay upon them a sheaf of freshly

....

recovered bodies to be removed to England, they were burned upon the beach on funeral pyres. In Dowden's recounting, there is no mention of whether Mary attended the cremations, no account of the sooty air. He tells us how long the bodies took to burn, how resistant they were to burning (the oft-told story of Percy's heart), how the sand, heated by the fires,

....

superposition—the idea that the deepest rock level is the oldest—and how it is, or can be, the opposite of *superimposition*. It is also one example of how my formal experiment here departs from geological concepts. The lowest band was not the first one written nor is it the oldest conceptually, although I did want grass to define the top and sand to define the bottom. Overall, though, I privilege superimposition and seek to superimpose temporalities—both in relation to the temporal gaps in my reading and in relation to the temporalities of the different layers here. I write these bands,

....

essay for this book, I thought (in a way that will sound far-fetched, if not absurd, but I hope to make convincing): those dots mark the space one must occupy and from which one must act in response to climate change.   °   Those dots, then, indicate a moment of hesitation, suspension, and deferral and invest it with value. But *A Room of One's Own*—initially delivered, in a shorter version, as two lectures to women at Girton

....

surfaces" and deep-time as "one and the same" (Sörlin, 276). As I stood and read this sign and thought that the rocks beneath my feet had been there for 440 million years, I felt time's fold, the sign signaling a bond we always have to what was here before, however remote.   °   But these bands are not

....

is as it were the interpretation of a diviner nature through our own; but

Layering 93

picked sweetgrass [. . . .] Will you hold the end of the bundle while I braid?

....

was "so scorched as to render standing on it painful" (Trelawney cited in Dowden, 577), and how the flames looked gorgeous in the evening light. ° 　 In a striking coincidence of fate, as many critics have noted, Shelley was working on a poem entitled "The Triumph of Life" at the time of his death. This title also chimes with the idea of *afterlife*. The afterlife,

....

then, as geological strata against the grain: lines that overlap even if they are spatially demarcated; stratifications that run not only straight ahead but also up and down and in reverse; palimpsests that draw out what has been written over. ° 　 My initial plan was to reread Derrida's essay from beginning to end, make notes; to reread Shelley's "The Triumph of Life"; and then to consult other critical interpretations, and compare and expand. As I proceeded, I realized that this method was what Derrida was, in part, referencing when he kept coming back to reading contexts while

....

College in October 1928—also charts the frustrations of interruption. One cannot think deeply if one is always interrupted. After the interruption above, the narrator retraces her steps and regains her path. She carries on thinking and arrives at Fernham at that time of day—"(seven twenty-three to be exact)" (16)—when the trees and butterflies and dust seem to gain an agency of their own. "It was the time between the lights," she continues,

....

like geological strata. The six bands do not resemble the sedimented layers of rock with the lowest layer signaling both the oldest datable period in time and the earth's origins. In short, my idea is imperfect. And so what is the point of this experiment? Is it just a clever trick—"a gimmick or a folly,"

....

its footsteps are like those of a wind over the sea, which the coming calm

Hands joined by grass, can we bend our heads together and make a braid

....

following Benjamin, is always incomplete and unfinished, never fully realized although, of course, not in the way that Percy's poem was, breaking off after the word "Of—," the dash defining the break. And yet it is that incompletion, that unfinishedness, that is life. The poem carries on, *lives on*, as Weber translates one meaning of afterlife (67). "The Triumph of

....

acknowledging that the contexts can never be saturated. There is always one more reading. And as I read, I remembered through Derrida's words that Shelley's poem is about ghosts and hallucinations, and that my readings and rereadings and more-readings would never settle into a singular reading. But they would give me enough of a handle on the contexts to begin, to walk into the text on a suspension bridge from which the far bank could not be seen. They would remind me that the indecision with which this text (and all texts) opens is not resolved by drawing lines between one

....

"when colours undergo their intensification and purples and golds burn in window-panes like the beat of an excitable heart; when for some reason the beauty of the world is revealed and yet soon to perish (here I pushed into the garden, for, unwisely, the door was left open and no beadles seemed about), the beauty of the world which is so soon to perish, has two edges, one of laughter, one of anguish, cutting the heart asunder" (16–17). She

....

to cite Bjornerud again—with little critical purchase? Further, how are you reading these bands? Are you reading all the bands on a single page? Or are you, like me as I read Derrida's bands, reading them in a linear fashion, one after the other? If so, how is this different from reading-as-usual (with an

....

erases, and whose traces remain only as on the wrinkled sand which paves to honour the earth? And then I'll hold it for you, while you braid, too"

....

Life" is the poem to which we turn for Percy's final thoughts. He had broken off from another work, a history play of Charles I, a month before his death. How different would our recollections of Percy be were they filtered through this history, which surely would also have been incomplete at the time of the fatal accident? Interestingly, Percy's reasons for setting the

....

band and another, introducing hyphens (small bridges) between one word and another, for it is precisely the line that connects and produces the superimposition; the band can always be opened, and new readings can always emerge from that opening, and there are always the ghosts of earlier texts hovering in the words we read, and what I wanted, I realized—the reason I had remembered Derrida's text as relevant to this book—was a way to read the ghosts (which is only another word for one type of mediation), and a way to move forward and to act in the context (infinitely saturated)

....

continues thinking and then is again interrupted, this time by the arrival of dinner captured by a dash: "For youth— [paragraph break] Here was my soup" (17). ° The narrator resumes her "search for truth" in the British Museum and is again waylaid. In this case, she stops herself in surprise when she recognizes that the vast number of works on women are written by men. "One went to the counter;" she writes, "one took a slip of

....

added irritation of turning more pages)? If not, how is there any sense on the page at all? ° In *Timefulness* Bjornerud writes that "rocks are not nouns but verbs—visible evidence of processes: a volcanic eruption, the accretion of a coral reef, the growth of a mountain belt. Everywhere one looks,

....

it. [. . .] it arrests the vanishing apparitions which haunt the interlunations

(Kimmerer, *Braiding Sweetgrass,* ix-x). "Now / make room in the mouth /

....

historical play aside were themselves bound up with reflections on time and its linear progression and, one could argue, these reflections involved a setting aside not only of the history of a king but of history itself in preference for the "passing moment." "I feel too little certainty of the future," he wrote his friend John Gisborne, "and too little satisfaction with regard

....

of indecision.   °   And yet: I didn't want abstraction; I wanted firm ground (not sand, not water); I wanted a place to stand; I didn't think—don't think—the world needs another theory book. And yet: How else to remind ourselves that we always act from somewhere along that figurative suspension bridge and that there is always a wobble in the mechanism? And yet: The "and yet" feels like a paw on the ground, preparing. And yet: Why return to an essay from November 10, 1977, to find tools for the present? I read Derrida's two bands exactly as I was hoping readers would *not*

....

paper; one opened a volume of the catalogue, and . . . . . the five dots here indicate five separate minutes of stupefaction, wonder, and bewilderment" (26). Here the dots open a space for astonishment at the extraordinary number of books men have written about women. They also mark minutes. Mary had imagined, she said, reading and then "transfer[ring] the truth [about women] to my notebook." But to do so, she realizes, she

....

rocks bear witness to events that unfolded over long stretches of time" (8). So, too, words, if across shorter timespans. In his 1865 book, *Words and Places*, Isaac Taylor writes that local place names are "records of the past" (1); they possess "great vitality," indicating "emigrations—immigrations—the

....

of life" (Shelley, "A Defence of Poetry"). "The sediments are a sort of epic for grassesgrassesgrasses" (Layli Long Soldier, *Whereas*, n.p.). "I believe a

....

to the past to undertake any project seriously or deeply. I stand, as it were, upon a precipice, which I have ascended with great, and cannot descend without greater, peril, and I am content if the heaven above one is calm for the passing moment" (cited in Dowden, 553). The past is a source of dissatisfaction and the future a source of uncertainty; Percy privileges instead

....

read my own bands: one by one. But I also realize, as I return to them now, that I do look at the entire page and realize something, probably the most obvious thing, and maybe I once knew this but I have forgotten it, that the upper band was a lecture that Derrida delivered. I realize now that the date noted above is not the date of that lecture but rather the date when Derrida dedicated his lecture to his friend, Jacques Ehrmann, who I surmise died sometime after so that the band below and the band above join together to ensure the living on of his friend who now lives on here

....

would need not to be human but rather a "herd of elephants" (for their long lives) or "a wilderness of spiders" (for their many eyes). She cannot possibly produce the comprehensive study of women and fiction that she had planned and will, instead, have to proceed more randomly. Those five dots may indicate stupefaction, but they also prompt a regrouping and a spur to further thinking. ○ While Woolf's book is in many ways

....

commingling of races by war and conquest, or by the peaceful processes of commerce:—the name of a district or of a town may speak to us of events which written history has failed to commemorate" (1–2). They may conserve "the more archaic forms of a living language or they embalm for us the

....

poem of the earth. [. . .] We might have expected the amount [of the leaf of grass is no less than the journey-work of the stars, / And the pismire

....

the solace of the passing moment and, perhaps, the poem on the triumph of life that is neither serious nor deep. °   Percy was writing his new poem at the same time as he was making himself a small boat of canvas and reeds. Before building the boat, he and Edward Williams outlined a plan for it "upon the sands of the Arno" (Dowden, 549). This sketch, this

....

too. °   Normally I would interrupt myself here to check these dates, learn a little more about the publication history, and likely find myself going off in new directions, following new leads, and experiencing a kind of delight that I always have when I do research. And yet such research, in the midst of climate crisis, feels like a luxury. I think of Benjamin in the internment camp in France, sitting on a bench, closed into himself and I think of the words of his biographers: *things were so bad that he could not read* (Eiland and Jennings 648). At any rate, I resist the urge to check such things.

....

about the obstacles to women's writing—the many interruptions that arrest our thinking, the many demands on our time—and the fact that English society does not carve out a place for women free from such interruptions, it is also about harnessing interruption to one's needs and becoming comfortable inhabiting that space of suspension. *Those three dots.* Mary wants to transfer the truth about women to her notebook. She is heartened

....

guise and fashion of speech in eras the most remote" (2). They are, as Adelene Buckland puts it, commenting on Taylor, "subject to the same natural forces as rocks" (155). Another geologist, Lauret Savoy, is alert to the relationship between place names and history. She cites Walt Whitman's claim

....

sediment blanket on the sea floor] to be vast, if we thought back through is equally perfect, and a grain of sand, and the egg of the wren" (Whitman,

....

writing upon the sand, materializes the "passing moment" to which Percy refers. Mary appreciates that boating made Percy happy; indeed, she writes, he composed much of "The Triumph of Life" while floating in the new canvas boat ("Editor's Note," 325). Percy carried the boat on his back like a turtle or hermit crab, Dowden writes, as if it were his

....

But I do interrupt myself to make some phone calls.   °   When I get back and lift up my book, several pages fall out. Bone-colored flecks of the binding glue are sprinkled across my desk like a light snow. Faded yellow post-its with ripped edges line the pages (I must have been trying to make post-its go further); the entire book, with its marginal notes, falling pages, and ragged borders, is a mess. Derrida is interested both in how we frame our objects of study and how they are mediated. As in indecision above, I feel as if there is something important to me here in relation to climate

....

when she gets to the nineteenth century publications and finds that women are finally writing. But they wrote mainly novels. Why? she wonders. "If a woman wrote, she would have to have a sitting-room. And, as Miss Nightingale was so vehemently to complain,—'women never have an half hour . . . that they can call their own'—she was always interrupted. Still it would be easier to write prose and fiction there than to write poetry or a play. Less

....

that "Names are magic. One word can pour such a flood though the soul"; they can, Savoy elaborates, "blaze with an intensity that seems to concentrate all life" (69). But she also cautions against a too-comfortable reading and seeks, instead, to ignite both what is there and what is not: "If history

....

the ages of gentle, unending fall—one sand grain at a time, one fragile

Leaves of Grass). "There are eggs of next summer's grasshoppers; there are

....

house.   °   Percy's mood of "unusual joy" that summer was in sharp contrast to Mary's (Dowden, 554). She wrote of this period many years later: "Sometimes the sunshine vanished when the scirocco raged—the 'ponente,' the wind was called on that shore. The gales and squalls, that hailed our first arrival, surrounded the bay with foam; the howling wind

....

change studies. The demand for narrative, Derrida notes, begins with the question: "'Tell us exactly what happened'" (87). This is the version of history that Benjamin, of course, rejects (Ranke's "the way it really was" [cited in Benjamin, "Theses," 255]). It is also the basis for what often counts as important in relation to climate change knowledge—a representation of what has happened and what we anticipate will happen next. It is around this point in the essay that Derrida tells us that his lecture did not seek to make things easy for its translators, but that the band below will try "for

....

concentration is required. Jane Austen wrote like that to the end of her days. [. . .] 'subject to all kinds of casual interruptions'" (70). The narrator later cites from *Jane Eyre*: "'When thus alone I not unfrequently hear Grace Poole's laugh. . . .' That is an awkward break, I thought. It is upsetting to come upon Grace Poole all of a sudden. The continuity is disturbed" (72). The three dots here signify not only, and not even especially,

....

can be read in the names on the land, then the text at the surface is partial and pieced. A reader might do well to look beyond 'official' maps for traces of other languages, other visions" (87). This process is also reflected in a different register in the most recent edition of the *Oxford Junior Dictionary*;

....

shell after another, here a shark's tooth, there a meteorite fragment—but the dormant seeds from which will come the grass, the herb, the oak tree"

....

swept round our exposed house, and the sea roared unremittingly, so that we almost fancied ourselves on board ship" ("Editor's Note," 323). Their house "almost stood amid the waves" (Dowden, 545). Mary writes that it was only when she was floating in Percy's boat that she felt content. The solitary location, "desolate house" (Dowden, 558), lack of comforts, worry

....

the greatest translatability possible" (89). It is also here that I have the urge, which I resist, to track the references to "in other words" (itself, of course, a definition of translation) in Derrida's essay. And it is here, too, that I notice that every time I turn the page, my eyes again go to the top of the page. This works when I'm reading the upper band, but it doesn't work when I'm reading the band below. It introduces a stutter or skid in my response, which I find interesting. It is an interruption that is so quickly corrected it hardly counts. And yet.   °   I continue reading, trying to

....

omitted content but rather an "awkward break" and a disturbance of the "the continuity." Sometimes the dots refer to minutes and sometimes they break with the minutes. They signal their own form. They remind us of mediation and teach us to read the mediations, in this context a mediation that interrupts and breaks linear time.   °   Mary concludes that without a space free from interruptions, the interruptions themselves will

....

Robert Macfarlane is dismayed to find "outdoor" words like "acorn, adder, ash, beech [. . .] willow" replaced with "indoor" words like "attachment, block-graph, blog [. . .] voice-email" (3). Words are like rocks and rocks are like words. "To think geologically," Bjornerud writes evocatively, "is to hold

....

the whole continuing persistently, relentlessly, endlessly. It is, of course, a

(Carson, *The Sea Around Us*, 36) "[We are] building a global, grassroots,

...

about her father, ill health, and a difficult miscarriage took their toll. ° Percy was not immune to these stresses. Shortly after Mary's miscarriage, he woke from an alarming dream: "Edward and Jane came in to him; they were in the most horrible condition—their bodies lacerated, their bones starting through their skin, the faces pale yet stained with

...

read, the upper and lower bands together. The upper band sidetracks to Blanchot's *Madness of the Day*, a text I know so little about that I do not even know if it is long or short. Again I resist the urge to interrupt myself and look up this text, although I am also struck by how different this interruption would be for me today than it would have been when I was a grad student in the 1980s. Then, I would go to the library, find the book in the stacks, check it out. Now, I would plug the title into Google and get, most

...

register in the form; there will be a "jerk" (72). The work "will be deformed and twisted" because the writer will be writing "in a rage" (72–73). Here I recall the two edges of the perishing beauty of the world—one of laughter and one of anguish—and marvel at how Woolf inserts her female writers, Grace Poole's laugh and Charlotte Bronte's rage, between their poles. And I marvel, too, that the perishing of our world incites a similar response of

...

in the mind's eye not only what is visible at the surface but also present in the subsurface, what has been and what will be" (22). Rocks, like words, are time-things. ° Rocks are also good for thinking. "No solid crystal." This is the way that Marx referred to industrial society in 1867. Society was

...

process similar to that which has built up layers of rock to make mountains" and broad-based network the likes of which the environmental movement

....

blood; they could hardly walk, but Edward was the weakest and Jane was supporting him. Edward said, 'Get up, Shelley; the sea is flooding the house, and it is all coming down'" (Dowden, 561). The dream, which woke Mary up, unsettled him for days, and it was in that period, too, that he saw the ghost of himself on the terrace. Just as the boat outlined on the

....

likely, multiple versions of the book. Online shopping would make it easy for me to order, or I could get it from my library, and it's not unlikely that an e-book would be available to me and I could have it on my desk in the time it would take to do these searches. But I don't do any of this. If I had, I would also most likely distract myself, notice certain writers I like who have commented on Blanchot's text and wonder what they have to say about it, notice others who have commented on Derrida's comments and

....

a double-inflected laughter (the laughter of glee and the laughter of grief) and a double-inflected anguish (the anguish of the loss of the ordinary day and the rage at the loss of the world).   °   But for all her meanderings, her losing her way, and her interruptions, Woolf does arrive at her point, a point that has been there all along. This point is the form. And this point is made through the novels of the fictional Mary Carmichael (a name that

....

not a crystal but "an organism capable of change" (Preface to *Capital*, 298). So, too, today: societies are organisms capable of change. But if crystals are solid, the word *crystal* is not. The word is derived from ice, and ice melts. Societies can disappear. The ice derivative follows from the fact that crystal

....

(Carson, *The Sea Around Us*, 76–77). "The strata of the Earth is a jumbled has rarely seen" (Klein, *This* 255). "We live in an age of rising seas. . . .

....

sand and the canvas boat itself brought into relief the liminality of land and sea, these events opened that liminal space between waking and sleeping and, more disturbingly, life and death. And, indeed, just as he planned his poem and his boat, he made plans for dying, requesting that Trelawney

....

want to know what they know, what they have written and, without quite articulating it to myself in this way, want to either "correct" or affirm my own readings. I love this part of scholarly work. It's a dialogue with the voices of others. But as Derrida's essay reminds me, the contexts are never saturated, and that tendency for textual freefall is only increased by the internet searches available. °   I turn to pages 100–101 and see two pages covered in my handwriting, a miniessay written in tiny penciled

....

folds back onto the fictional narrator ["call me Mary Beton, Mary Seton, Mary Carmichael or by any name you please" {5}] and speaks not only to collaborations between women but also to collaborations across time). For Carmichael, the interruptions do not mar but rather enable the writing. Woolf's narrator takes Carmichael's book from "the very end of the shelf" and considers that, while it is the author's first book, "one must read it as

....

is transparent and refractive; we see through it, and it breaks up light. Surely one of the reasons that Benjamin was drawn to the metaphor of crystal was for its mediation of light. Lyell writes of the rocks he observes: "This process cannot have been carried on for an indefinite time, for in that case all the

....

museum" (Smithson, 28). "To work in the elements of the Earth, in

I have wondered at the feeling of the sea that came to me [. . .] wondered

....

buy him enough prussic acid to take his life should "an irrevocable malady" occur (Dowden, 555). The prussic acid was like a line in the sand that might result in the construction of a sturdy boat or might be washed away. I don't think that Percy planned to die, but he said to Marianne Hunt

....

words in the margins of the page. Here Derrida turns to "superimprinting" and "the logic of the 'double bind'" (101), a logic that his two bands clearly perform in this essay. I pause and turn the page. Derrida returns to Shelley's unfinished poem and his drowning. His comments on unfinishedness are worth citing here, but I carry on. Derrida uses Lydia Davis's translation of Blanchot's *Death Sentence*. "As defined (indefinitely) in the passage from *Le pas au-delà*," he writes, "the *arrêt de mort* is not only the decision

....

if it were the last volume in a fairly long series, continuing all the books I have been glancing at [. . .] . For books continue each other despite our habit of judging them separately" (84). The way in which the first and the last are interwoven here and the books are set into a conversation across time at once respects their historical positioning and sets them into a broader conversation (as, indeed, Woolf has been doing throughout this

....

stratified rocks would long ere this have been fused and crystallized. It is therefore probable that the whole planet once consisted of these mysterious and curiously bedded formations at a time when the volcanic fire had not yet been brought into activity. Since that period there seems to have been a

....

the materials fetched from geological timescales, was to seek refuge for until I remembered that the hard rocky floor on which I stood, its flatness

....

(Leigh Hunt's wife), on the day before the boating accident: "If I die tomorrow, I have lived to be older than my father. I am ninety years old" (Gibson, 228).   °   *I am ninety years old.* Percy's sense of time's elasticity, its extension through experience and its compression in rocks and

....

that determines [*arrêtant*] what cannot be decided: it also arrests death by suspending it, interrupting it, deferring it with a 'start' [*sursaut*], the startling starting over, and starting on, of living on. But then what suspends or holds back death is the very thing that gives it all its power of undecidability—another false name, rather than a pseudonym, for difference. And this is the pulse of the 'word' *arrêt*, the arrhythmic pulsation of its syntax in the expression *arrêt de mort*. *Arrêter*, in the sense of suspending, is suspending

....

lecture).   °   I pause here because I have, as I often do, a sense of what I want to write but an inability—a stutter and a skid—to quite articulate it. Instead I look out my window. It is an unseasonably warm February day, although the fields on the farm where I write are still filled with snow. I think of how two of the works on which I am focusing in this section were first lectures and later published in print form. I am also struck by the

....

gradual development of heat; and this augmentation we may expect to continue till the whole globe shall be in a state of fluidity and incandescence" (*Principles* [e-book], 69). He was writing in the first half of the nineteenth century before thinkers began to reflect on the consequences not of "a

....

the human in the nonhuman. What will remain are layers, only layers" interrupted by upthrust masses of jagged coral rock, had been only recently

....

words and sand, is captured in this phrase. And yet, despite the prussic acid and the premonitions, Percy surely felt he had more time. Despite some scholars' speculations, he surely did not set out in that sailboat to make, in Mary's words, a "plaything" of his fate ("Editor's Note," 324).

....

the *arrêt*, in the sense of decision. *Arrêter*, in the sense of deciding, is arresting the *arrêt*, in the sense of suspension. They are ahead of or lag behind one another. One marks delay; the other, haste" (114–15). This language, goes to the heart of what I want to do with post-time. And later: "The indecision of the *arrêt intervenes* not *between* two senses of the word *arrêt* but *within* each sense, so to speak. For the suspensive *arrêt* is *already* undecided *because it suspends*, and the decisive *arrêt* undecided because what it

....

coincidence of suicides, and I wonder if, reflecting on ecocide, my own reading might be unconsciously gravitating to suicide. And then I delete those words because they feel too blunt and clumsy. And then I write them again. Ladybugs line my window. They lift their small wings and try to fly but hop instead, making a soft sound. I look back to Woolf's narrator's comments on Mary Carmichael: "To begin with, I ran my eye up and

....

gradual development of heat" but of a heart-stoppingly rapid development of heat. Whether we may expect a global "fluidity and incandescence" or something else remains an open question.   °   Crystals, rocks, stones, earth are written into our language; they are foundational, so to speak, and

....

(Sörlin, 44). He "went down onto the sloping sand and slipped among constructed by the busy architects of the coral reefs under a warm sea. Now

....

When he figuratively stood upon that precipice, he seemed to seek not peril but calm. A wife and young son awaited him back at Casa Magni along with an unfinished poem entitled "The Triumph of Life." Unlike Mary, it is doubtful he reflected on the dangers he courted. And yet it is

....

decides, death, *la Chose*, the neuter, is the undecidable itself, installed by decision in its undecidability" (115). In stopping, the title (*Death Sentence / L'arrêt de mort*) of Blanchot's novel sets things in motion and its unreadability is what makes reading possible.   °   And then the essay veers into death. Or rather, it has been *about* death for many pages, but Derrida now begins to intertwine *Death Sentence* with the double binds of language, demands, and writing—the "death sentence" becomes a sentence

....

down the page. I am going to get the hang of the sentences first [. . .] . So I tried a sentence or two on my tongue. Soon it was obvious that something was not quite in order. The smooth gliding of sentence after sentence was interrupted." (84) She continues: "Something tore, something scratched; a single word here and there flashed its torch in my eyes. She was 'unhanding' herself as they say in old plays. [. . .] to read this writing

....

this written foundation, far from being sturdy and stable, both crumbles and is uncontained, unbounded by precise temporal markers. I want these bands both to remind us of geological striations, the sedimentation of meaning, its sandiness, and to perform a sort of destratification. I've tried to

....

the currents, which quickly immersed him" (Blanchot cited in Derrida,

Layering 109

the rock is thinly covered with grass and water; but everywhere is the feeling

....

hard not to see how many unfinished things he left in his wake.   °   As Percy, Edward, and Charles left the harbor at Lerici on Monday, July 8, Captain Roberts had binoculars trained upon their boat from a lighthouse to which he had gained access. He watched it sail alongside other boats, he

....

about death that becomes a double bind: "If you don't kill me, you'll kill me" (cited on 118). I can think of many other interpretations beyond Derrida's, but I respect the point he is making, that the translator, by deciding on this translation rather than that one, always leaves something out, and that translation is only an extreme and vivid example of how we always communicate. The double bind can also be the beauty of holding incommensurable things together. Still, I am approaching the midpoint of

....

was like being out to sea in an open boat. Up one went, down one sank" (84). It is dizzying and disorienting, and suddenly we are in another boat: Woolf, Carmichael, and her characters are all together "in a canoe up a river" (85).   °   The next paragraph continues: "I am almost sure, I said to myself, that Mary Carmichael is playing a trick on us. For I feel as one feels on a switchback railway when the car, instead of sinking, as one has

....

do this in many ways—some obvious, and some perhaps less so, some that work and some that probably do not. I think here of Woolf's comment: it is as if I am in a cave lit by a candle, unable to see where I step.   °   These bands are linear and layered on the page, but this banding together is also

....

"Living On," 82). "It happened one day, about noon, going towards my that the land has formed only the thinnest veneer over the underlying

....

watched the gathering storm on the horizon, the boats then small dots upon the sea, and he watched as the boats disappeared into the "haze of the storm" (cited in Dowden, 568). Only Percy's boat did not emerge when the storm passed. From the harbor Edward Trelawney similarly watched

....

Derrida's essay, and I sense my desire to slow down, to immerse myself in the words, to stop my deliberate skimming (which is also conducive to skidding), and to read slowly, to return to the beginning of the sentence or word, or quotation, and to read again. To pause. To transcribe and make notes (holdover from my grad student days: I still sometimes write down notes on a pad of paper that is to my right, but these moments are reserved for difficult passages, when I want the self-imposed discipline of slowing

....

been lead to expect, swerves up again. Mary is tampering with the expected sequence. First she broke the sentence; now she has broken the sequence. Very well, she has every right to do both these things if she does them not for the sake of breaking, but for the sake of creating. [. . .] And determined to do my duty by her as a reader if she would do her duty by me as writer, I turned the page and read . . . I am sorry to break off so

....

meant to be a braiding together. If the braiding does not happen, the bands may still signal the role of mediation in how we communicate, but the place of connection and collaboration will be lost. Needless to say, I did not want simply to add an annoying element to the reading experience that offered

....

boat, I was exceedingly surprised with the print of a man's naked foot platform of the sea, that at any moment the process might be reversed and

....

anxiously for Percy's "boat amongst the small craft scattered about. I watched every speck that loomed on the horizon" (cited in Dowden, 569). Before the storm, he writes, "the sun was obscured by mists" and the air "was oppressively sultry." "The sea was the colour, and looked as solid and

....

down, the barely discernible scratch of the pen on paper).   °   I turn the page. I am guided now by what my twenty-something self found important: my eyes linger on the sentences I have underlined, the words and phrases in the margins, the post-its. I know, because this has happened before, that what I saw then is likely not the same thing as what I want to pay attention to now, but this time I am trying another approach. Derrida's text, with its line dividing the upper band from the band below, invites my

....

abruptly. Are there no men present?" (86). Those three dots famously capture Mary's shock when she reads that "'Chloe liked Olivia. . . .'" Three more dots, dots that also disturb the continuity. "Do not start. Do not blush," she continues. It happens: "Sometimes women do like women" (86). We have been on a boat in an open sea, heaving up and down. We have been on a railway, swerving when we thought we might plummet.

....

little by way of critical interest. Worse, while I am exploring the role of interruption in this book, I am wary of forms of interruption that interrupt sustained thinking. I want to cultivate the pause *for* thinking that Benjamin valued in interruption. But how does one do that? Benjamin's work is

....

on the shore, which was very plain to be seen in the sand: I stood like the sea reclaim its own" (Carson, *The Sea Around Us*, 97, 100). "Its surface is

....

as smooth as a sheet of lead, and covered with an oily scum. Gusts of wind scuffed over without ruffling it, and big drops of rain fell on its surface, rebounding, as if they could not penetrate it" (cited in Dowden, 568). Soon the sky darkened and the "fury of the storm" broke overhead, the

....

attention to my own penciled underlines. Lines, that unlike the printer's line, have a slight wobble, a slight tremor as if they were written on a suspension bridge. On one page this is the only thing I have underlined: "He is told to 'Come,' *and* she's dead" (121). I realize that I am reading a shadow text, a text within the text, carved out by an earlier version of myself, resurrected now in this new reading in the context of climate change. Blanchot writes about death, defeat, fighting death, not getting help when things

....

And then we arrive at the three dots that signal something new, that speak what has not been spoken before: Chloe liked Olivia. . . . °   I am sorry to break off so abruptly. This writing is not a lecture, and I cannot gauge my audience or ask permission. . . . I want, cautiously, hesitantly, to think about the climate crisis in the space of those three dots. Cautiously, hesitantly, because, of course, Woolf's book is not about climate

....

helpful in this context. He turns to form. Like Woolf, he considers how form, especially surprising or difficult forms, can produce a pause that makes us think and notice how form itself works. I am hoping that seeing these forms on the page makes you hesitate. And perhaps, occasionally, look

....

one thunder-struck, or as if I had seen an apparition" (Defoe, *Robinson* covered with grass and rushes, presenting a dry crust and a fair appearance;

....

crashes of the thunder and sound of the rain erased all human noises. It lasted only about twenty minutes.   °   Of all the accounts of the premonitions of that day, Mary's are perhaps the most poignant. She is aware, of course, of the pitfalls of reading retrospective premonitions into this

....

might still change, a shift in direction when it is "too late," when death becomes inevitable, a moment when all one can do is grieve (122–23). And the band below these reflections invokes "apocalypse" (125). The text begins to breathe.   °   The narrator of Blanchot's text, Derrida notes, repeatedly says that he cannot say (127). So too does Derrida. Derrida puts it this way: "He is forbidden to say. So—he says" (127). He says the hyphen, the dash, the suspension bridge that speaks only in the tremor. The forbidden

....

disruption. Or is it? I want to consider what it means to be lost in thought, take a different path, act from the uncertainty there. I want to transpose "stupefaction, wonder, and bewilderment" from the books that are not on the shelves (women's) to the feet that are not on the streets (most people's). I want to multiply the "awkward break[s]" in whatever way enables us to hear what we have not yet heard. Like Woolf, I don't think this action will

....

up and down the page to see what other bands occupy the same space. The bands do not point to an origin as geological rock sedimentations do. Indeed, if they do anything along those lines, they gesture away from origins. There are six beginnings to this chapter (that follow the three beginnings to

....

Crusoe, 184). "*How the sun burnt uncompromising, undeniable. It struck*

but it shakes under the least pressure, the bottom being unsound and

....

period of Percy's "last days." But she does so anyway. She describes Percy's pleasure in boating and the ways in which he— and all of them—ignored the dangers posed by the sea. As "a child may sport with a lighted stick," Mary writes, "till a spark inflames a forest and spreads destruction overall,

....

becomes the bidden, or rather, the bidden inhabits the forbidden and barely speaks, is forbidden/bidden to speak. These are the microreadings I have been avoiding, forbidding myself, in my reading of Derrida's essay thus far, but here, in almost the exact center of the text, I find them puncturing my writing, unbidden.   °   The essay is after a "'logic' that would enable us to read *everything* [. . .] down to the smallest element, the grain of sand, the letter, the space. . . . Everything that, in the text above, goes

....

come from one person. It will come from all of us acting together. I don't know what form it will take or where it will come from but I want to make a call—a "peroration," to follow Woolf (114)—for creating the conditions for its emergence. Woolf was surprised by the swaying and sinking, the turning and swerving of the open boat of Carmichael's writing; it startled her and commanded her attention. She looked up. She looked down at the

....

this book). Chakrabarty, among others, argues that climate disruption calls for new ways of thinking: "The crisis of climate change calls for thinking simultaneously on both registers [capital and species], to mix together the immiscible chronologies of capital and species history. This combination,

....

upon the hard sand, and the rocks became furnaces of red heat; it searched semifluid. The adventurous passenger, therefore, who sometimes in dry

....

so did we fearlessly and blindly tamper with danger" ("Editor's Note," 324). Children lack the experience to understand the consequences of playing with fire, and Percy, like a child (a comparison also reinforced by many of his biographers), succumbed to the pleasures of his environment.

....

back to the dissemination of sand (beach, seaside, hour-glass). The temptation to translate (turn over, transfer) Blanchot's hour-glass into Shelley's '. . . and whose hour/ Was drained to its last sand in weal or woe, / So that the trunk survived both fruit & flour.' '. . . And suddenly my brain became as sand. . . .'" (122–23). The ellipses are sprinkled like sand in Derrida's text. I read them too.   °   Later, we return to the after-life, which seems to be a living on that is also a death sentence. Which itself is the title,

....

page and read some more. She began that dialogue with the last woman writer's first novel and, in doing so, gestured toward those who would follow. "So much," Woolf's narrator writes, "has been left out, unattempted" (86). We, too, can look up, look down, feel the rise, the almost-sink, and use those three dots as a path, an invitation, an interruption that makes a place for the unattempted.   °   For Woolf, such a venture is not

....

however, stretches, in quite fundamental ways, the very idea of historical understanding" ("Climate of History" 220). To "mix together" capital and species history involves forging a different approach to history that departs from earlier models. To mix, or braid, together ideas hitherto kept separate

....

each pool and caught the minnow hiding in the cranny, and showed the rusty seasons traverses this perilous waste, to save a few miles, picks his cautious

....

But Mary found no pleasure there: "During the whole of our stay at Lerici, an intense presentiment of coming evil brooded over my mind, and covered this beautiful place, and genial summer, with the shadow of coming misery—I had vainly struggled with these emotions—they seemed

....

untranslatable and so mistranslated (like all translations) here. And yet this phrase, unlike other mistranslations, signals a series of truth slippages that amount to a knowledge, unsayable, that is above and below truth—the two bands that are the double bind—and are only truth. After which Derrida turns to beginnings and ending and, specifically, the end, the second part, of Blanchot's text. Another undecidable line, the line between night and day. I read: "The truth beyond truth of living on: the middle of the *récit*, its

....

without risks. Those who express something new "will light a torch in that vast chamber where nobody has yet been. It is all half lights and profound shadows like those serpentine caves where one goes with a candle peering up and down, not knowing where one is stepping" (88). I take heart in the fact that the sexism Woolf describes is now vastly, if imperfectly, improved.

....

in the Global North demands new ways of thinking. ° This call for new ways of thinking, I am suggesting, is enabled by new *forms* of thinking. It also invites us to engage with what we sometimes think of as modes of layering that defy or baffle temporality. It invites us to consider, for

....

cartwheel, the white bone, or the boot without laces stuck, black as iron in way over the rushy tussocks as they appear before him, for here the soil is

....

accounted for by my illness, but at this hour they recurred with renewed violence.... . I could scarcely bring myself to let them go" ("Editor's Note," 324). Mary interleaves beauty with "the presentiment of coming evil," "the shadow of coming misery," the "whisper[] of coming disaster" (324). The

....

element, its ridge, its backbone [*arête*]. There is only one blank space in the typography of the book" (144). And my handwritten note in the margin: "like Heid's threshold." (I read this first as "like skid's threshold" because my handwriting is unclear.) I puzzle over this for a moment, no longer remembering exactly what Heidegger meant by the threshold but knowing that I was absorbed for many years with Heidegger until a boyfriend brought that absorption to a halt, indeed a skidding halt, with his casual

....

But this imperfect improvement has been many, many centuries in the making. It is perhaps willful, perhaps reckless, to compare Woolf's call for women to pick up their pens and begin to write with a call for all of us, in whatever ways we are able, to think about and respond to climate change. But Woolf many times beseeches us to look one hundred years forward

....

example, Shelley's dream that "the sea is flooding the house, and it is all coming down," Woolf's observation that "it is in our idleness, in our dreams, that the submerged truth sometimes comes to the top" (*A Room*, 31–32), George Eliot's brief elegy to lost idleness in her comments on

....

the sand" (Woolf, *The Waves*, 105 [italics in original]). "As archaeology of firmest. If his foot slip, or if he venture to desert this mark of security, it

....

"prognostics" may be "inaudible," but they are "not unfelt" (324). And from all of this emerged not the raging forest fire intimated above but a sense of "the unreal": "The beauty of the place seemed unearthly in its excess: the distance we were from all signs of civilization, the sea at our

....

comment, "How can you so uncritically endorse the thinking of a Nazi?" ° My eyes slide down the page, and I contemplate citing a long passage—it continues for several pages in the slender slip of the band below—on institutions like the university, the distinctions they enforce between departments, no doubt much diluted since Derrida's writing but still deep-rooted, and on writing, reading, and transference. ° The upper band returns to *Hamlet* with Derrida's observation that the "*récit*'s the thing" (145). It is

....

from the time that she is writing and to see a transformed world. She even describes that world and, while we are not quite there, it is comparable to our own: the "nursemaid will heave coal. The shop-woman will drive an engine" (40). Women now do jobs that were once assigned only to men. But look again at the jobs: heaving coal, driving an engine. We have to put

....

post-time, and my own forgetfulness with respect to Derrida's essay. Our idleness, our dreams, make possible a power that bends thoughts and words and forms, a potency, a palimpsestic thinking, not always available when our lives are imagined as lines. °   As rocks are layered—or, better,

....

our thought easily shows, man is an invention of recent date. And one is possible he may never more be heard of" (Lyell, *Principles* [e- edition]

....

feet, its murmurs and its roarings in our ears,—all these things led the mind to brood over strange thoughts, and, lifting it from every-day life, caused it to be familiar with the unreal" (324). The last word in Mary's account is "wrecked" (325).   °   (While critical commentaries focus on

....

the thing—and the thing in this essay is death, not understood as death in terms familiar to us but rather death as living on, death as after-life—because it neither presents nor represents a truth (the "tell us exactly what happened" that runs through this essay, along with "this time" and "in other words," like leitmotifs) but performs that truth.   °   And now Derrida returns to ideas of unfinishedness—how can we ever say where this text ends? A long section follows in which there are many quotations from

....

pressure not on who is doing the jobs but rather the jobs themselves. And the systems they support. The terms have shifted. One hundred years from when Woolf wrote would be 2028. Our decade. I like to think that if Woolf walked among us (and by her theory, she does), she would fashion her words to address the climate crisis. It would not take

....

layering—and thick with historical sediment, so too, figuratively, are words. Consider the word "band." It derives from the Old English *band* that was linked to another Old English word, *binden*. *Band* is bound to *bind* as we can see in usages such as a rubber band that ties things together, a headband

....

perhaps nearing its end. If those arrangements were to disappear as they

724). *Gramineæ*: "the order of plants to which grass belongs" derived from

....

how those at Casa Magni, and Percy himself, saw Percy as a ghost in those weeks before his death, it is Mary, after Percy's death, who I most think of as a ghost, not a ghost through which one walks or sees, but a ghost into which one bumps. A ghost that does not yield. A ghost that is as hard and

....

Blanchot's text that I only half-follow. A marriage proposal is made but not in the language of the person to whom it is made. She is lost in the crowd. I flip the pages, feeling that these pages are not as important to me (and knowing, too, that such an impression may signal that they are). And suddenly there are two women and I know I have been reading too quickly, because where did this other woman come from?—"interruption" (164). On the following page temporality words like "while," "as soon as,"

....

many adjustments.   °   I write this book in a room of my own. The interruptions that distract me from thinking are different from the interruptions that distracted women in Woolf's period but, as she notes, "interruptions there will always be" (81). Some interruptions are enabling and some are disabling, and, again to cite Woolf, "All of this [how to harness

....

that contains hair, a band that branches out into belts, ties, cords, and ribbons. *Band* also suggests *stripe*—a striped band on a sock—which generates *stria*, *striation*, and *lane*. The variations on *striation* bring us back to geology. But I paused over *lane*. A lane keeps things contained, and it reminded me

....

appeared, if some event which of the moment we can do more than sense *gramen* meaning grasses. The suffix -gram (ideogram, telegram, Instagram)

....

lasting as stone: "On that terrific evening [I was told] I looked more like a ghost than a woman; light seemed to emanate from my features, my face was very white, I looked like marble" [cited in Dowden, 572].) °   The omens, premonitions, and pleas not to travel take on a chilling cast in

....

"immediately," and so on are piled up. An interdiction happens "in the quasi-middle" of the text, and this middleness makes me alert to mediation (a word Derrida never uses to my knowledge). And suddenly, it seems, this is a text about interruption.   °   Of course when one writes on something, one sees it everywhere, and, as a literary scholar, I am adept in making a case for the relevance of what has been overlooked or what seems inconsequential. (I pause over both of these words too.) Derrida, I realize, is

....

interruption productively, how to limit interruptions that deplete one] should be discussed and discovered" (81). While the ". . ." is generally left to signify loosely, there is one moment in the last chapter of *A Room of One's Own* in which Woolf develops the "lull and suspension" to which the three dots gesture: "At this moment, as so often happens in London, there

....

that bands are also about containment and staying in one's lane. In this context, the other branch of *band* is useful to me. Derived from the Old French *banden*, "band" also means banner and is the origin of our understanding of musical bands. Bands, in this sense, suggest many musical

....

the possibility—without knowing either what its form will be or what derives from the Greek *gramma* indicating "thing written, letter of the

...

light of what follows. We fold our knowledge back and see the month from mid-June to mid-July in 1822 not as an unremarkable month but as Percy's "last days." Why did Percy not heed the warnings? When I read about this lonely house by the sea, the waves lapping at its walls, the wild

...

putting women in two different Blanchot stories into dialogue with each other (an approach he recognizes as "mad" [170]). I reflect briefly on the fact that this is exactly what I am doing in this chapter: putting into conversation, if not directly, voices from different books and different periods. Derrida notes that the "interruption between the author and the narrator, or indeed between the two women" is not "simple"; "it is as ambiguous as the interruption of every *arrêt de mort*" (170). And then the text shifts to

...

was a complete lull and suspension of traffic. [. . .] A single leaf detached itself from the plane tree at the end of the street, and in that pause and suspension fell. Somehow it was like a signal falling, a signal pointing to a force in things which one had overlooked. It seemed to point to a river, which flowed past, invisibly, round the corner, down the street, and took

...

instruments playing together. The musical band also reminds us that each individual band is *already* composed of many parts playing together, as is, indeed, each word. The marching band perhaps brings both of these meanings together, as a group plays in a musical band but also stays in its lane. I

...

it promises—were to cause them to crumble, as the ground of Classical alphabet" (OED). "In the fields with which we are concerned, knowledge

....

weather, and the coming storm, I interpose not only my reading of what I know followed for the Shelleys but also my knowledge of sea waters rising and unparalleled storms around the world. We don't know what follows for us. But we do know more than Mary knew when she pleaded with

....

the mediation, registered by the dash, of the "tele-phone" and the "telegraphic" (170). All of this is repeated with a slight variation immediately after this passage (171). And then: "I must break off here, interrupt all this, close the parenthesis, and let the movement continue without me, take off again, or stop, arrest itself, after I simply note this:" (172). I remember now the disappointment I experienced at the conclusion of this text, because it felt *unfinished*, because the author (Derrida) really does break off after what

....

people and eddied them along, as the stream at Oxbridge had taken the undergraduate in his boat and the dead leaves" (100). This passage revels in the lull and suspension, the pause, the fall and the flow, the eddies, and the force in things which one had overlooked. ° Let us consider Woolf's words again: "the beauty of the world is revealed and yet soon to perish

....

am hoping that the bands here will be "played together" but that they will also serve as banners proclaiming a position. In other words, I want both the band that dissolves singular meanings and the band that maintains them. The ". . ." marks the space of that possibility. I also want to be alert

....

thought did, at the end of the eighteenth century, then one can certainly comes only in lightning flashes. The text is the long roll of thunder that

....

Percy not to go. We know about storms and what they portend. °   We know about planetary boundaries. We read all accounts in retrospect and prospect. I think now of the rising sea as a signal from centuries ago, a beacon to our own times, like Greta Thunberg's warning, *Our house is on fire*, and hear, superimposed on the premonition of the fatal boating

....

he says, and he instead cites a long passage from Blanchot (and "The Triumph of Life" does not come into his reading in any extended way at all)—the passage runs for three pages—and then the essay stops. °   Or does it? Because I have forgotten, or rather, set aside—for it is staring at me the entire time—the band below. I have a new eagerness as things, in relation to this essay, are coming to an end, for I know I can return to these final pages of the band below and, perhaps, my disappointment will be eased. For now: there is a death mask, and I think of Benjamin's death

....

[. . .] , the beauty of the world which is so soon to perish, has two edges, one of laughter, one of anguish, cutting the heart asunder" (16–17). This is a description of the last light of the day, soon to fade to dark. On October 2, 1940, Woolf wrote a diary entry in which she imagined what it would be like to be killed by a bomb: "Terrifying. I suppose so—Then a swoon; a

....

to the unruly. I want to encourage a moving into other lanes, driving off the road, getting lost, and the new sounds—the new banners—that follow from that. The bands in this chapter, to be sure, are also driving off one version of the academic road, and while they may sometimes collide and crash,

....

wager that man would be erased, like a face drawn in sand at the edge of follows" (Benjamin, *Arcades*, 456).

....

accident, the sea that is flooding our own house. *The sea is flooding our house*, Percy cried out in his dream. This knowledge may feel as if it comes to us in a dream, but it bears listening to just as the inhabitants listened to the waves crashing on the shore close to Casa Magni "like that of booming artillery."

....

mask in relation to allegory. But, no, I read the band below and continue to feel—disappointed. But my disappointment ignites something: the appointed time. *Hamlet* again. The ghosts it invokes. *Bleak House* and its reference to ghosts, *Hamlet*, and appointed times. And the realization that to be disappointed is, perhaps, to have my "points" disintegrated, discollected, scattered across the ground, to borrow a metaphor from Derrida, "like sand" (122).

....

drum; two or three gulps attempting consciousness—& then, dot dot dot." Six months later, on March 28, 1941, Woolf left her house in the morning, walked down to the River Ouse, filled the pockets of her coat with rocks, and . . .

....

and they may sometimes spin out of control, I hope they also sometimes make their own lane, or better, enable us to get out off the lane or road all together and explore—"discuss and discover," in Woolf's words—new paths hitherto untraveled.

....

the sea" (Foucault, *The Order of Things*, 387).

On Time: Second Experiment

In the Idiom of the Self-Help Guide

In *Middlemarch*, George Eliot famously offers an astute and penetrating anatomy of procrastination through the story of Edward Casaubon. With the Industrial Revolution and its paeans to productivity well underway, the concept of procrastination was gaining new prominence in her period.[1] The term itself does not come up in nineteenth-century titles related to writing; instead, it was typically used to reference delay in the context of religion (sealing one's relationship with God before death) and, somewhat oddly, delay in the economic context of insurance (buying an insurance policy before fire hits, for example).[2] But the practice of procrastination, and the implications of that practice, were already being introduced into middle-class culture in the context of a society increasingly drawn to what we might now call productivity. In 1859, just a decade before the publication of *Middlemarch*, for example, Samuel Smiles's *Self-Help* offers myriad strategies for overcoming procrastination—while not once using the term itself.[3] Eliot's treatment of Casaubon, in this context, can be read as a reverse mirror image to Smiles's almost comically uplifting case studies.

The reverse mirror image also extends to the form of Eliot's case study.[4] *Self-Help* is composed of thirteen chapters further subdivided into sections ranging from a paragraph to a few pages. Its fragmented form, repetitions, potted histories, aphorisms, and 750 microbiographies stand in sharp contrast to *Middlemarch*'s narrative complexity.[5] It may, accordingly, seem misguided to distort Eliot's sensitive account of psychology and productivity into a series of "lessons" as I do below. And it may seem even more misguided to extend those lessons to the climate crisis. What can *Middlemarch* have to say about either of these things in the register of self-help? And yet, as Leah Price records, Eliot's work has a long history of being treated for its pithy remarks and wise sayings, one that Eliot encouraged not only through her own reliance on epigraph and quotation but also through the maxims her narrators issue across her work. Alexander Main's *Wise, Witty, and Tender Sayings in Prose and Verse Selected from the Works of George Eliot* coincides, as Price notes, with the part publications of *Middlemarch*. Its warm reception led to updated editions and birthday books routinely issued over the following decades (106).[6] Further, Price argues, Eliot likely anticipated such extractions and wrote with their possibility in mind.[7] That said, to reduce *Middlemarch* to a series of self-help mantras sits uneasily not only with the multifaceted complexity of the novel itself but also with the larger goals of my book.

Dorothea, an ardent twenty-year-old woman, marries Casaubon, an older man who is both a clergyman and a scholar working on a book that he hopes will, to put it colloquially, change the world. The marriage is not a success. He dies with the book unfinished. Many more things happen in *Middlemarch* but this is the crux of the novel for my focus. Casaubon never quite masters his struggles with what we would now call procrastination. Can these struggles teach us something about the climate crisis, the "book" of which we are all still collectively writing so to speak?[8] Or, to ask this question in way that more closely aligns with my goals: Can we distill transportable lessons from *Middlemarch* while remaining mindful of the fact that nothing transports without difference, relation, friction, and pier-glass-like refraction and distortion?

If these obstacles are not grounds enough for caution, a further methodological difficulty also obtains: the limitations of situating what is a collective problem—the climate crisis—in the flamboyantly individual context of self-help. There may perhaps be some fruitful anachronistic points of intersection between Casaubon's case and the extraordinary number of recent self-help books dedicated to "fixing" procrastination. But these points of intersection are surely less pertinent when it comes to the world action reform demanded by the climate crisis. As I indicate elsewhere in this book, moreover, the idea of the atomistic self, not to mention the extractivist methodology I deploy here, underpins the larger systemic models that have contributed to the climate crisis in the first place. Nevertheless,

I will develop below a number of implicit "Procrastination Lessons" articulated in *Middlemarch* that match in very close terms the procrastination lessons discussed in self-help literature since the early 1990s.[9] In other words, I will read *Middlemarch* against its stated grain. These lessons will be coupled with "Climate Action Lessons" that focus variously on the collective and the personal. This chapter, accordingly, has a self-help tone built into it which risks taking the climate crisis too lightly. But I am interested in holding open performative contradictions—preserving and canceling both the line and the singular subject, for example—and this experimental chapter is written in that context.

1. PROCRASTINATION LESSON #1: *BE WARY OF ASPIRING TO GREATNESS*.

Climate Action Lesson #1: *Be wary of trying to solve the climate crisis.*

In *Middlemarch*, Casaubon is famously hesitant and hopeful with respect to his "great work" (14), the "Key to all Mythologies." The title does not mince his ambition. The reader is told very early on that this key to *all* mythologies is his life-long dream, that he labors over it assiduously and daily, and yet that it remains unfinished. The first chapters of the novel trace, from Dorothea's point of view, the shift from her warm embrace of Casaubon's project to her slow disillusionment. In many ways, she has married Casaubon for the promise of his book and what she can learn from it: she imagines that the book will improve the world, add to its knowledge, and she wants to dedicate her own life to helping him realize his goal. She is "captivated by the wide embrace of [his] conception" (14) and eagerly supports "the highest purposes of truth" by which he is driven. Unstinting in her own pursuit of "the good of all" (9), Dorothea may be forgiven for attributing the same generous attributes to her future husband. The narrator describes Casaubon's work-in-progress through Dorothea's eyes as follows:

> Dorothea by this time had looked deep into the ungauged reservoir of Mr. Casaubon's mind, seeing reflected there in vague labyrinthine extension every quality she herself brought; had opened much of her own experience to him, and had understood from him the scope of his great work, also of attractively labyrinthine extent. . . . With something of the archangelic manner he told her [Dorothea] how he had undertaken to show (what indeed had been attempted before, but not with that thoroughness, justice

of comparison, and effectiveness of arrangement at which Mr. Casaubon aimed) that all the mythical systems or erratic mythical fragments in the world were corruptions of a tradition originally revealed. (14)

Earlier Casaubon described himself to Dorothea as a kind of "ghost" who "lives too much with the dead" (Dorothea is enchanted), bent, with weak eyes, over ancient books and "confusing" material (9). Here he is compared to an archangel. But the time-bending transports of ghosts and angels I have discussed in the previous chapters are not in evidence here. Instead we encounter, Procrastination Lesson #1: Be wary of aspiring to greatness. This procrastination lesson will arise again and again in the pages of *Middlemarch* as Casaubon's aspirations for critical acclaim coupled with his self-doubt interfere with his ability to complete his work to his satisfaction. The goal for which Casaubon "aimed"—not only the "thoroughness," "justice," and "effectiveness" of approach but also the conception of a singular key—is too high.

Aspiring to greatness, counterintuitively, can also impede climate action. Such aspirations can be especially problematic when one seeks solutions.[10] First, like Casaubon, it is easy for individuals, organizations, and countries alike to be daunted by the "great scope" of the task at hand. Like Casaubon's projected "Key," the work is complex, multidimensional, and demanding. That said, Casaubon's conception of his project is misguided from the start.[11] The effort to reduce carbon emissions is not. Nevertheless, in both cases, success will be better ensured by approaching the problems presented with an open and flexible mind divested from ego-driven motivations. In Casaubon's case, his aspirations to greatness make him reluctant to include others in his deliberations. In the case of climate action, aspirations to greatness can lead to competition between countries or vested interests to the exclusion of the collaborative work required.

2. PROCRASTINATION LESSON #2: *DIVIDE YOUR PROJECT INTO DOABLE CHUNKS*.

Climate Action Lesson #2: *Divide your activism into doable chunks.*

The "attractively labyrinthine extent" of Casaubon's project above is a red flag. Dorothea doesn't see it at first, but the reader does. Contra what Dorothea admires in Casaubon, to avoid procrastination one should aspire to a project that is tightly gauged, that is not vague, that is not a labyrinth, and that similarly has an easily graspable extension whose scope is neither too far nor too uncertain. Casaubon's focus on the end result or what he calls the "crowning task," moreover, only compounds his already dismaying breadth problems. The passage cited above continues as follows:

> But to gather in this great harvest of truth ["Procrastination Lesson #1" again] was not light or speedy work. His notes already made a formidable range of volumes but the crowning task would be to condense these voluminous still-accumulating results and bring them, like the earlier vintage of Hippocratic books, to fit a little shelf. (14)

Casaubon seems here to be back in line with basic procrastination lessons: he recognizes that things worth doing are rarely "light or speedy work," and to accomplish them one must divide the work into doable chunks—say, Casaubon's volumes of notes. This is all to the good. But embedded within this description is a deal-breaking flaw, one that unravels—or, rather, harnesses in the service of procrastination—the very procrastination prevention that Casaubon's approach seems to articulate here. To recognize the flaw in his approach, it is helpful to turn to Casaubon's advice to his cousin, Will:

> "I have insisted to him [Will]," [he tells Dorothea] "on what Aristotle has stated with admirable brevity, that for the achievement of any work regarded as an end there must be a prior exercise of many energies or acquired facilities of a secondary order, demanding patience. I have pointed to my own manuscript volumes, which represent the toil of years preparatory to a work not yet accomplished." (55)

Here two problems emerge that are also evident in the earlier quotation: the focus on "an end" and the focus on "a prior exercise of many energies," "patience," "the toil of years," and "preparatory" work all capped by "not yet accomplished." Let's look again at the earlier quotation: however valuable it is to produce volumes of notes, the "still-accumulating results" chime with Casaubon's endorsement of years of toil and preparatory work that can, if not properly calibrated, eclipse the end itself. The "crowning task," like the "end," is, in fact, impeding Casaubon's ability to accomplish his goal. Instead of dividing his work into doable chunks toward the realization of a realistic and well-conceived goal, he aspires too high (Procrastination Lesson #1). And—the point of this procrastination lesson—instead of closing off his work in doable chunks, he continues to accumulate results and unspool preparatory work without clearly defined boundaries. Worse, he uses this conviction that years of unharnessed toil are essential to his project to procrastinate its completion further.

This procrastination lesson maps onto climate action as follows: preparatory work must be done, and that work is often slow and laborious. But individuals and collectives alike need to transition from this work to action when the preparatory work is completed to one's satisfaction. Perhaps even more important in the context of climate action, the work must be divided into doable chunks both at the preparation and the action stage. Making it too big and focusing too intently on the "crowning task" will only thwart success. Just as it is impossible to sit down and write a book like "The Key to all Mythologies" all at once so, too, it is impossible to realize the reduction of carbon emissions all at once. Both tasks are too big. But they can both be broken down into doable chunks, the end result of which will approximate the goal at hand. In each case, the goals may at any moment seem susceptible to Casaubon's error of endless accumulation. Which brings us to Procrastination and Climate Action Lesson #3.

3. PROCRASTINATION LESSON #3: *KNOW WHEN TO STOP TAKING NOTES AND START WRITING.*

Climate Action Lesson #3: *Know when to stop gathering information and start action.*

Casaubon appears incapable of transitioning from the note-taking stage to the hard work of synthesis and writing. The problem with writing projects is that the preparatory work is never finished. Every student writing a dissertation, and every academic writing a book, knows this. There is always one more book or article or chapter to read; and that book or article or chapter always alerts one to still more in a never-ending series.[12] If one has "already made a formidable range of volumes" of notes, then one should be at the writing stage. Taken together, Casaubon's "still-accumulating results" and his focus on the "crowning task," projected into an ever-receding distance, read as a veritable procrastinator's guide to what not to do.[13]

A similar problem besets climate change action in the Global North. The "still-accumulating results" of climate science steadily proliferate as goals are set (the "crowning task" of carbon emissions reduction) while the division of the work into doable chunks lags behind. The Global North, too, has a formidable range of volumes stretching from Intergovernmental Panel on Climate Change reports through scientific analysis and extending, more recently, into climate change research in the social sciences and humanities. But the climate action equivalent of a complete book continues to be deferred. Of course successful action requires "years of toil" and "preparatory work," but at some point the emphasis has to shift from the generation of information to its application in action items. Instead of emissions decreasing to the degree required by that formidable range of trusted expert volumes, they are on the rise or declining so negligibly as hardly to count at all.

4. PROCRASTINATION LESSON #4: *BE WARY OF SELF-CONSTRUCTED DELAYS AND DISTRACTIONS*.

Climate Action Lesson #4: *Be wary of self-constructed as well as collective-constructed (nations, NGOs, organizations, and corporations) delays.*

Casaubon has been working on his project for thirty years (refer to Procrastination Lesson #3) when he proposes to Dorothea. This proposal displaces his great, labyrinthine, and voluminously note-filled work. The narrator writes: "Mr Casaubon, as might be expected, spent a great deal of his time at the Grange in these weeks [before the wedding], and the hindrance which courtship occasioned to the progress of his great work—the Key to All Mythologies—naturally made him look forward the more eagerly to the happy termination of courtship" (41). But—and this suggestion may at first seem like a stretch—perhaps he proposes to introduce a seemingly valid procrastination device; perhaps, in other words, he proposes *to procrastinate*? Perhaps the surprising choice to marry at his advanced age is just the procrastination obstacle he needs to ease his frustration with his great work?

If marriage is a device to procrastinate finishing his project, in Casaubon's case it is not wisely chosen. Dorothea does not want to impede

Casaubon from his "great work"; on the contrary, she wants to advance it. Like Casaubon, she is eager "to arrive at the core of things" (42). The "core of things" in which this novel is most interested is the personal. "Suppose we turn from outside estimates of a man," the narrator writes, "to wonder, with keener interest, what is the report of his own consciousness about his doings or capacity: with what hindrances he is carrying on his daily labour; what fading of hopes, or what fixity of self-delusion the years are marking within him; and with what spirit he wrestles against universal pressure, which will one day be too heavy for him, and bring his heart to its final pause" (56–57). We all feel pressures external to us to complete certain projects or get work done. But the internal pressures, the narrator reminds us, exert their own distinctive anguish. And they can lead into further self-deluding actions.

Like Casaubon's "Key," work to address climate change has been taking place over the past thirty years. Here, too, a proposal may be afoot, although what form that proposal takes remains to be seen. It may be a proposal to Big Oil, to green energy companies, to geoengineering companies, or to countries and nations collectively. Like Casaubon, it may be issued with the best of intentions, and the Global North may tell itself that it can defer no longer and that, in fact, the proposal will aid the important climate action work it seeks to accomplish. The world will be lucky if the proposal is to some idea that embodies the values that Dorothea promotes.

What we do know is that the self-constructed delay of a proposal and a honeymoon backfires for Casaubon. He imagines that he's bringing a wife into his life who will support his procrastinatory strategies. He is wrong. The next procrastination lesson illustrates just how wrong he is.

5. PROCRASTINATION LESSON #5: *DO NOT BE AFRAID TO ACT NOW*.

Climate Action Lesson #5: *Do not be afraid to act now.*

Chapter 20 opens with Dorothea on her honeymoon, crying alone in her room in Rome. It is six weeks after her wedding. Casaubon is doing research at the Vatican, and she is distressed by an inchoate sense that this was not the wedding trip she imagined, Casaubon is not the man she imagined, and Rome is not the place she imagined. Indeed the chapter dilates—one could perhaps say procrastinates—for six dense pages before turning to the cause of Dorothea's tears. It unfolds that Dorothea is crying precisely because she and Casaubon have had a dispute about procrastination.[14] I will follow the narrator's lead, however, by first attending to the dilatory material that offers context on Dorothea's state of mind—her own "core of things"—before turning to their exchange.

From the outset, Dorothea imagines she will learn great things from Casaubon. While there are some suggestions that she may be misguided in this hope, the fullness of her error comes in Rome. It slowly dawns on her that she will not likely learn *in general* from Casaubon. This recognition is followed by the harsh revelation that she will also not learn *in particular* from Casaubon, that perhaps Casaubon's "Key to all Mythologies" is lacking in range and richness of vision, that it is not grounded in adequate scholarship, and that Casaubon's mind is ominously "weighted with unpublished matter" (138).[15]

The structure of this procrastinatory interlude is as interesting to me as the "pause" discussed in *Adam Bede* in "Post-time." It is an effort to capture Dorothea's interiority, but it is still inflected very much by the narrator's voice. This section begins, for example, "Yet Dorothea had no distinctly shapen grievance that she could state even to herself" (133). The challenge for the narrative is: How does one represent that "confused thought" that one cannot even represent to one's self (133)? The narrator does so primarily through Dorothea's response to Rome. Its "stupendous fragmentariness" and its "ruins"—a word that is used repeatedly—bewilder Dorothea in a manner that parallels her response to her marriage. The narrator reflects on the fact that Casaubon is still Casaubon, but Dorothea's perspective on him has shifted: as the narrator puts it, "the light had changed" (135).

In the middle of this section that relays Dorothea's turbulent response to Rome and her increasing disenchantment with her marriage, the narrator places what many critics agree is the most famous line in the novel: "If we had a keen vision of all ordinary human life, it would be like hearing the grass grow and the squirrel's heart beat, and we should die of that roar which lies on the other side of silence" (135). Alexander Welsh reads this passage as an elaboration of "what cannot be known,"[16] and we could see it in tension, then, with Casaubon's "Key" and its aspiration to tabulate and organize all knowledge. But I want to consider what it might mean to place this passage in the procrastinatory pause that delays relating why Dorothea is now, as the narrator puts it in the same paragraph, "in a fit of weeping six weeks after her wedding" (135). The narrator explains just before this passage that a little crying is not unusual and she does not think that readers will be deeply moved by it. But if we were, in fact, really attuned to "ordinary human life," it would be too much to bear. What these lines do, in this procrastinatory pause, is place us in an inflationary present.

But it is excessive. We must move on. That "roar which lies on the other side of silence" is too much to "bear." Dorothea is living in her "new real future" instead of in the future she imagined;[17] her life has not been put off, she is in it, and it is hard. But it *is* life. And it gives a woman's weeping a sustained value coupled with the quest to understand and articulate the causes of this weeping. This section is part of Eliot's development of psychological interiority: she values weeping, she values the twenty-year-old just-married woman, she values the things that make her weep. And the procrastinatory pause lets her foreground this value and its relationship to interiority as it also offers a gentle reminder that life is lived in the present. However energetically one might dwell on past issues or project a better future, the "real future" is always now. In that "real future" that is the present, pace post-time, one lives out the consequences of previous actions, and that living is embodied; it includes weeping and the heartbeats of small things and the sound of grass growing.

When the narrative finally refers back to (or arrives at) the argument about procrastination that prompted Dorothea's tears, Dorothea and Casaubon are having their morning coffee. Casaubon says to Dorothea, innocently enough, "We must now think of all that is yet left undone, as a preliminary to our departure. I would fain have returned home earlier . . . ; but my inquiries here have been protracted beyond their anticipated period" (138). Entirely unaware of Dorothea's state of mind, he introduces two related ideas that will have a bearing on Dorothea's response: things "left undone" and more work that remains to be done. He continues: "I have been led farther than I had foreseen, and various subjects for annotation have presented themselves which, though I have no direct need of them, I could not pretermit" (138). Here we have a reminder and restating of Procrastination Lesson #3: Know when to stop. Be wary of times when inquiries extend too far beyond their anticipated period. If one has "no direct need" for material, do pretermit. Especially on one's honeymoon.

In response, Dorothea presents herself as the procrastinator's enemy in a passionate passage in which she hopes to be helpful while dispensing with all of Casaubon's procrastinatory strategies in one fell swoop. She says: "'And all your notes. . . . All those rows of volumes—will you not now do what you speak of?—will you not make up your mind what part of them you will use, and begin to write the book which will make your vast knowledge useful to the world?" (139). To make matters worse, she ends "with a slight sob and eyes full of tears" (139). She follows Procrastination Lessons #1 through 3: she does not refer to his work as great, she does not refer to his work as large, and she suggests that he stop the note taking *now*. The more material one has, the more difficult it is to know what to use and what to discard. For a hardcore procrastinator, however, it would be the words between the dashes that would be most difficult to countenance: *Will you not now do what you speak of?*

Climate action is often oriented toward the distant future (figured as apocalypse or catastrophe rather than a written work). But like Dorothea, we all live only in a "new real future," an inflationary present, which is now. The new real future of the now does not often realize the projections of the imagined future (soon to be the "new real future" as time passes), it can be difficult to bear, it can be overwhelming, but it is where the work happens. Further, the still-accumulating data can be overwhelming to process. Dorothea's simple point is to return to the project we have charted for ourselves and to ask: Are we now doing what we have spoken of? Or are we hesitating, fearful and uncertain about how to proceed?

6. PROCRASTINATION LESSON #6: *DO NOT KEEP YOUR WORK TO YOURSELF; SHARE IT WITH OTHERS, EVEN "IGNORANT ONLOOKERS."*

Climate Action Lesson #6: *Do not work in isolation; talk to others, share ideas. (This climate lesson has one important caveat: Do not share your work with the ignorant onlookers who are climate deniers. There is little point and they will likely be a distraction. Just do the work with the 97% of like-minded others who share your views.)*

Casaubon is predictably upset to hear Dorothea's tear-filled appeal to get the work done. His response to Dorothea matches his inner irritation: "My love . . . you may rely upon me for knowing the times and the seasons, adapted to the different stages of a work which is not to be measured by the facile conjectures of ignorant onlookers" (139–40). He does not want to aim for the small achievement, which would be easy enough; he does not want to be judged by critics who do not understand his subject matter. Here he flouts Procrastination Lesson #1—the greatness of his project—and articulates the new procrastination lesson above.[18]

Dorothea, however, once again brings Casaubon right back to what he wants to avoid and asserts, in the process, her capacity to judge *this one thing*, not the quality of the project itself, but the fact that he is not finishing it. She says: "You showed me the rows of note-books—you have often spoken of them—you have often said that they wanted digesting. But I never heard you speak of the writing that is to be published. Those were very simple facts, and my judgment went no further" (140). Casaubon again confronts the terrible recognition that Dorothea will not be the aider and abettor of his procrastination but rather someone capable of prodding him on to action and "agitating him cruelly." "Instead of getting a soft fence against the cold, shadowy, unapplausive audience of his life," he wonders, "had he only given it a more substantial presence?" (140). Procrastination has allowed Casaubon to live in the twilight zone of working but not really working. In the absence of a "substantial presence" reminding one otherwise, one can always, pace Procrastination Lesson #3, feel that more work is required, one more visit to the Vatican or the library, one more tractatus to write or article to read, one more notebook to fill. In a contemporary context, Mark Kingwell writes of this all-too-familiar academic impulse always to read just one more article: "This potentially constant deferral, so comforting and pervasive, is a delicious foretaste of the entire conceptual universe of humanities scholarship, so often in the grip of what Gilbert Adair called, in his essay "Derrida Didn't Come," the 'future procrastinate'" (369–70).

In this lesson, I often think of Casaubon as a condensed, singular version of all the global leaders confronted by the rising, sometimes tear-filled, certainly passionate, cries of the young school strikers around the world. You've showed us the science, they cry out, you have your evidence. Will you not now do the work that is so necessary for the survival of us all? Why do you delay? And these global leaders, agitated cruelly, all too often reply with a voice as if issued from lofty heights: You are young and you are in no position to evaluate. And the rejoinder of the school strikers: Maybe so, but we do know this one thing: you are delaying. In all your years of research and weighing options, carbon emissions have not improved in any appreciable way, and by some measures—your measures, based on the science you've relayed—they have got worse. You have rows and rows of data, but we have never heard you speak of what is to be done exactly, what the precise plan is. These are very simple facts, and our judgment goes no further.

7. PROCRASTINATION LESSON #7: *BE WARY OF THE PROCRASTINATOR'S DILEMMA*.

Climate Action Lesson #7: *Be wary of the push and pull between doubt and grandeur in one's actions.*

The first line of Chapter 29 introduces a startling narrative interruption that has commanded the attention of critics: "One morning, some weeks after her arrival at Lowick, Dorothea—but why always Dorothea?" (192). Once again this is a passage that has been interpreted in many ways; I want to focus, however, on the way that this interruption foregrounds a distinct self with a distinct point of view. Procrastination, unlike narrative delay, requires a procrastinator, a self that procrastinates. And this self, in the nineteenth century, is also typically a self that doubts. Indeed, when the novel turns to Casaubon's point of view in this passage—which like the passage from *Adam Bede* could be understood, following Caroline Levine, as a "self-reflexive interlude" (*Serious Pleasures of Suspense*, 104)—his profile as procrastinator is immediately reinforced: he procrastinates marriage and he procrastinates writing, the narrator reminds us. Casaubon had done "nothing exceptional" in getting married; he had only felt that he should "defer" matrimony no longer. In one of the many very funny passages in this novel, the narrator relays Casaubon's reflections on marriage in terms of leaving "behind a copy of himself which seemed so urgently required of a man" by sonneteers of the sixteenth century. But "times had altered since then" and now "no sonneteer had insisted on Mr Casaubon's leaving a copy of himself; moreover, he had not yet succeeded in issuing copies of his mythological key" (192). Here "the book" is associated with children, and the narrator implicitly introduces yet another procrastinatory strategy, following Procrastination Lesson #4, oft-discussed in the procrastination literature today: having a child.[19] But Dorothea has no desire to slow down Casaubon's intellectual production with the distraction of children. She wants to speed it up. And this is, again, where Dorothea and Casaubon collide.

The narrator notes that it was not Casaubon's fault that he expected to be happy in marriage. But he had innate shortcomings that followed from a lack of passion. His "soul . . . was too languid to thrill out of self-consciousness into passionate delight; it went on fluttering in the swampy ground where it was hatched, thinking of its wings and never flying" (193). This is as good a description as any of Casaubon's procrastination of his book (not to mention his possible impotence).[20] In these contexts surely there is merit in Casaubon's doubts and fears that his work might not be good enough. He is that psychological type that has been much anatomized by the twentieth- and twenty-first-century procrastination theorists: that person who has illusions of grandeur but who is also wracked by the deepest insecurity.[21] This pull in two competing directions is the procrastinator's dilemma, a dynamic that is not lost on Dorothea. Following Will's troubling claim that Casaubon is not equipped to address his subject because he has neglected to learn German, Dorothea reflects "what could be sadder than so much ardent labour all in vain?" (154).[22]

Of course "so much ardent labour all in vain" is precisely what Casaubon most fears. For he too has an inkling, if not a full knowledge, that his work may not realize his grand ambitions. This inward fear is matched by a fear of the approbation of others: of the scholars in his field, of Dorothea, of Will. Casaubon wants his ideal imagined self to be known and applauded; but he "fears most of all" that it will not be. He feels confident that his "conduct" is "unimpeachable" (193), but his confidence wavers when he thinks of the work that most matters to him, his "Key to All Mythologies":

> The difficulty of making his Key to all Mythologies unimpeachable weighed like a lead upon his mind; and the pamphlets—or "Parerga" as he called them—by which he tested his public and deposited small monumental records of his march, were far from having been seen in all their significance. He suspected the Archdeacon of not having read them; he was in painful doubts as to what was really thought of them by the leading minds of Brasenose. (193)[23]

Most critics focus on Casaubon's obsession with his reputation, and he does throughout the novel exhibit a sensitively strung susceptibility to the scholarly opinions of others.[24] But these are also the painful reflections of the procrastinating writer who, pace Procrastination Lesson #1, aspires too high. The effort to write the "unimpeachable" (read: impossible) work produces instead the very "small" if monumental (I wondered, indeed, what "small monumental" was meant to signify here) pamphlets, the marginal pieces, which go unread or unappreciated while the bigger project remains undone.[25] To aspire to the unimpeachable, in other words, is to hamper one's best efforts. These reflections are capped by a dramatic illustration of the person who aspires too high. Above all, do not make your project a condition for immortality as Casaubon does: "The consolations of the Christian hope in immortality seemed to lean on the immortality of the still unwritten Key to All Mythologies" (193). In short, torn between the hope that his "great work" will be his ticket to immortality and his "painful doubt" that it will not, Casaubon embodies the procrastinator's dilemma.

Climate action, of course, is often beset by a similar dilemma: Is the work worth doing or is it useless? This question arises especially intensely in relation to the actions of individuals, thus lodging doubts in those who seek to solve the climate crisis (grandeur) but are uncertain what their action means and fear being scorned by others for their misguided attempts (doubt).[26] As with writing projects, the best antidote against this obstacle is a generosity of response. A person's or a collective's value cannot be tied to the success of endeavors pursued. They will always fall short. And the very standards of evaluation are constantly shifting. In short, reject the terms of the procrastinator's dilemma. Neither writing projects nor climate action should be reducible to personal or collective value. All of us, writers and activists alike, need the embrace of kind and supportive environments, the sort, indeed, in which most humans learned to walk and talk: environments where failure and falling down is part of the experiment of learning.[27]

8. PROCRASTINATION LESSON #8: *BE WARY OF REQUIRING IDEAL CONDITIONS IN WHICH TO WRITE.*

Climate Action Lesson #8: *Be wary of requiring ideal conditions in which to work.*

The narrator ends the commentary on Casaubon's fears with respect to reception with one of the more biting comments in the entire novel: "For my part I am very sorry for him" (193). Pity! Many pages have been dedicated to demonstrating that Casaubon is a character painfully sensitive to the pity of others. This pity, moreover, is related to precisely that result of procrastination that so troubles both philosophers and psychologists: the way in which procrastination creates a limbo zone that preempts living itself. The narrator writes:

> It is an uneasy lot at best, to be what we call highly taught and yet not to enjoy: to be present at this great spectacle of life and never to be liberated from the small hungry shivering self—never to be fully possessed by the glory we behold, never to have our consciousness rapturously transformed into the vividness of a thought, the ardour of a passion, the energy of an action, but always to be scholarly and uninspired, ambitious and timid, scrupulous and dimsighted. (193)

And yet who among us has not—to adopt the narrator's tone—at some point felt scholarly and uninspired, ambitious and timid, if not also scrupulous and dimsighted? The modifier "always" suggests the extremity of the case, but this particular set of characteristics is likely not that unusual.

Still, for all of these reasons—because Dorothea refuses to countenance the greatness of the project, because she wants Casaubon to get it done, because she is her own substantial self—Casaubon wants to "defer" bringing Dorothea into his study to help him. But she insists. She either reads aloud to him or copies out work:

> The work had been easier to define because Mr Casaubon had adopted an immediate intention: there was to be a new Parergon, a small monograph on some lately-traced indications concerning the Egyptian mysteries whereby certain assertions of Warburton's could be corrected. References were extensive even here, but not altogether shoreless; and sentences were actually to be written in the shape wherein they would be scanned by Brasenose and a less formidable posterity. (194)

The "not altogether shoreless" is promising. Indeed, here in a rare moment, Casaubon affords a positive example of one of the procrastination lessons, Procrastination Lesson #2: Divide the project into doable chunks. This period of writing, perhaps because he is finally heeding a procrastination lesson, is one of Casaubon's "busiest epochs" (194) marred only by intimations of a visit, quickly quashed, from his cousin. He explains the declined visit to Dorothea as follows: "I trust I may be excused for desiring an interval of complete freedom from such distractions as have been hitherto inevitable, and especially from guests whose desultory vivacity makes their presence a fatigue" (195). And here he gives voice to Procrastination Lesson #8: Be wary of requiring ideal conditions in which to write. The condition that is perhaps most often resorted to is Casaubon's here—the need for "complete freedom" and an absence of distractions to do one's work—but there are many others.[28] Dorothea speaks sharply in turn, and shortly after their heated exchange Casaubon, agitated, has a "fit" (196). Dorothea is alerted to the problem "when she heard the loud bang of a book on the floor" (196); this image is only marginally less subtle than Leonard Bast's death-by-bookcase in *Howards End*.

While Casaubon seeks ideal circumstances for working free from distractions, these rarely arise in ordinary life. Similarly, with respect to climate action, there is always a figurative Will knocking on the door and interrupting one's work. But the work must go on even so. Waiting until oil companies are better positioned for energy transitions, waiting until voting constituencies are more open to climate action, waiting until budgets can sustain, and networks support, the changes desired only will ensure that the necessary shifts will never be realized. There will always be reasons to do nothing, but Climate Action Lesson #8 reminds us to look more closely and to be wary of waiting for ideal conditions.

By this point in Eliot's novel, the narrator's treatment of Casaubon has helped to articulate eight procrastination lessons in the context of which I have articulated eight climate action lessons. His "fit in the library" (196) marks the beginning of Casaubon's steady decline, and while there is much more to be addressed in terms of his procrastination, I will leave his case study here.

9. PROCRASTINATION LESSON #9: *DO WRITE AS IF YOU MIGHT DIE TOMORROW*.

Climate Action Lesson #9: *Do act on climate change as if you might die tomorrow (with caveats)*.

It is probably clear by now that the self-help dimension of this chapter—the procrastination and climate action lessons—tracks what I have called the *procrastinator's dilemma*: Is this a scholarly approach or is it simply another superficial contribution to the 68,500,000 plus articles and websites on procrastination? Does it have any intellectual merit or is it trivial? And in the case of this book, does the way in which this chapter contradicts both what comes before and what comes after undermine the book as a whole? These are questions to paralyze the writer prone to procrastination.

Nevertheless, risking—or embracing—the trivial, I want to close with two final procrastination lessons. They may be interpreted as the positive partners—the "Dos" to the negative "Do nots" and "Be wary ofs"—already outlined. In the middle of the night, shortly before his death the following day, Casaubon wakes Dorothea, and together they work on his "Key to all Mythologies."[29] She marks passages at his dictation, exactly the process of sifting that Casaubon had suggested, but deferred doing, earlier. They seem to get a lot done even if Casaubon is not in fact writing. Indeed, Casaubon has been warming up to completion since his earlier "fit" above. His "mind was more alert, and he seemed to anticipate what was coming after a very slight verbal indication, saying, 'That will do—mark that'—or 'Pass on to the next head—I omit the second excursus on Crete.' Dorothea was amazed to think of the bird-like speed with which his mind was surveying the ground where it had been creeping for years" (330). Often the intimation of death is the one thing powerful enough to shake the fear of working and propel one forward. Or, stated more positively, in terms more relevant to most of our lives: do work with a fixed deadline.

In the context of the climate crisis, this lesson takes on a different cast. Many climate activists feel, like Casaubon, as if a death sentence has been delivered. But unlike the death sentence for the individual, the death sentence for the human species rearranges all of the lessons outlined above. Casaubon is spurred to complete his project, in part, because he dreams of his work's immortality. But immortality looks very different in the context of the human species's demise. Indeed, it can as easily be a disincentive as a prompt to action. And so for climate activists, this climate action lesson has to be carefully navigated. The deadlines—in the form of dates—issued by the Intergovernmental Panel on Climate Change and other governing bodies are important as context and catalysts for action. At the same time, because the *as if* overlaps so closely with the *is* here—work *as if* you are going to die tomorrow often translates in climate action into *you are going to die tomorrow*—it can serve as an impediment to action. All of these points only make the last lesson more important.

10. PROCRASTINATION LESSON #10: *MAKE IT SOCIAL; WORK WITH A BUDDY, BUILD A COMMUNITY*.

Climate Action Lesson #10: *Make it social; work with a buddy, build a community.*

Casaubon finally draws Dorothea into his work, albeit at a time when she no longer believes in it and no longer wants to be there, and he benefits from their shared comradeship (330).

This same advice to make it social and work with others holds for climate change action and, indeed, is the lesson that has underpinned all of the lessons above.[30]

BONUS LESSON:
PROCRASTINATION AND CLIMATE ACTION LESSON #11: KNOW WHEN TO TAKE A BREAK.

And especially: know when to relax and pick up a book. If the book you pick up is *Middlemarch*, you will also discover yet another point that is apt for this self-help guide: there are no rules, or lessons, that are absolutely applicable all the time.

. . .

On Time: Third Experiment

Found Questions

> *In what follows we shall be questioning. . . . Questioning builds a way.*
> —MARTIN HEIDEGGER

In this book I have been exploring what happens when "ways of meaning" (Weber, 75) come to the fore. In my first beginning, "Interruption," I cited Weber's comment that "translatability" gestures toward meaning without reaching it, that it "constitutes a *way* . . . rather than a *what*" (92). Translatability, which can be counterpoised to the focus on the *what* (or what happened) in realist novels, documentary, and historiography, only offers an extreme example of how all meaning works. But how can more vagueness, rather than more precision, help a climate crisis already beset by doubt and uncertainty? The point, I think, with Weber's focus on "the way," is that however much one wants to deny or walk away from the precarity of meaning and, indeed, the vulnerability of living, it is always there. Better, then, to confront it directly—to look into and stay with the gaping hole, as it were—and craft responses alert to its implications. This chapter foregrounds "the way" via questions. It is a "found chapter" composed of questions excised from their contexts and put into conversation with other questions. Inspired by Henry Mayhew's and Walter Benjamin's shared interest in ragpicking and collecting, and Benjamin's example of juxtaposing quotations without commentary, I bring together questions found in a range of works that speak in varying registers to the climate crisis.

There is, of course, a long tradition of question and response in Western philosophy. From Plato through Hegel, Marx, Arendt, and Heidegger, questions manifest thinking. Judith Butler writes that the tradition of philosophical inquiry on which Derrida, for example, relies "took the question as the most honest and arduous form of thought" ("Jacques Derrida"; see also Derrida, *Specters of Marx*, xix). In a social and political context, conversations, dialogues, and interviews form one end of a continuum, the other end of which is interrogations and inquisitions. Questions are shaped by their forums of exchange, in other words, and these are not always kind or conducive to further thinking. For Heidegger, though, questions are a form of way making or path making (although the political questions he himself failed to ask are a sobering part of his legacy). His translator, William Lovitt, aptly figures his work as "sandcastles" (Lovitt, "Introduction," xxxviii). Here I seek to hold open a space in which questions collide and rub up against each other in what I hope will be productive tensions, if not, also, provisional paths or sandcastles for thinking.

The questions here are not chosen evenly or fairly, and some of the questions are repeated. I was not careful to include a range of positions or views. Rather I chose the poetic and the provocative. While some of the questions are taken from works of poetry, most are not, and some fall on the page like roadblocks one has to climb over. Some questions sustain argument while also interrupting it. Some are incisive critiques, some are thoughtful inquiries, some are efforts to get closer to a point through rephrasing, some repeat for emphasis, some are impatient, some are rhetorical, some are frustrated, and some are hopeful. All are open-ended invitations to further questions.

· · ·

There was a strange stillness. The birds, for example—where had they gone?[1]
"What happened to the frogs? We don't hear them calling anymore."[2]
How did it come to this?[3]
"Who has given anyone the right to cut down trees and destroy a habitat for the sake of a double-page advertisement for cars?"[4]

Cheap gas now or maples for the next generation?[5]

What does consumer choice mean compared against 100,000 years of ecological catastrophe? What does one life mean in the face of mass death or the collapse of global civilization? How do we make meaningful decisions in the shadow of our inevitable end?[6]

We are . . . driven to the conclusion, that causes generally quite inappreciable by us determine whether a given species shall be abundant or scanty in numbers. Why, then, should we feel astonishment if the rarity is occasionally carried a step farther,—to extinction?[7]

Because the earth has reabsorbed the dead into its elements for so many millions of years, who can any longer tell the difference between receptacle and contents?[8]

. . .

But where have we strayed to?[9]

How did it come to this?[10]

How did we get caught up in this mess?[11]

What is wrong with us? . . . What is wrong with us?[12]

And what is this? / Whose shape is that within the car? & why . . . / is all here amiss?[13]

How do we interrupt the perpetual circuits of fear, aggression, crisis, and reaction that continually prod us to ever more intense levels of manic despair?[14]

What kinds of human disturbance can life on earth bear?[15]

What does it mean to hug the bear?[16]

Will he ever see a moose? . . . Will he ever see a bat?[17]

"What is it like to be a bat?"[18]

How does a bat hug?[19]

What does it mean to know but not to grasp, to have realization end in a shrug?[20]

[Thought] swayed, minute after minute, hither and thither among the reflections and the weeds, letting the water lift it and sink it, until—you know the little tug—the sudden conglomeration of an idea at the end of one's line: and then the cautious hauling of it in, and the careful laying of it out?[21]

We all think we know and we all think everybody knows. But we don't. Because how could we?[22]

What explains our ability to separate what we know from what we believe, to put aside the things that seem too painful to accept? How is it possible, when presented with overwhelming evidence, even the evidence of our own eyes, that we can deliberately ignore something—while being entirely aware that this is what we are doing?[23]

. . .

One touch of nature may make the whole world kin, but usually, when we say nature, do we mean ourselves?[24]

It may be true, as a mystic once contended, that most people, sometime in their lives, are moved by natural beauty to a "mood of heightened consciousness" in which "each blade of grass seems fierce with meaning," but the question is: What meaning? "All nature," contended another mystic a century ago, "is the language in which God expresses his thought." Very well, but what thought is that?[25]

"Why do rivers flow? Why does rain fall? Why does the sun warm us? And the wind, why does it blow?"[26]

This green flowery rock-built earth, the trees, the mountains, rivers, many-sounding seas;—that great deep sea of azure that swims overhead; the winds sweeping through it; the black cloud fashioning itself together, now pouring out fire, now hail and rain; what *is* it? Ay, what?[27]

How would the rest of nature respond if it were suddenly relieved of the relentless pressures we heap on it and our fellow organisms? How soon would, or could, the climate return to where it was before we fired up all our engines? How long would it take to recover lost ground and restore Eden to the way it must have gleamed and smelled the day before Adam, or *Homo habilis*, appeared? Could nature ever obliterate all our traces? How would it undo our monumental cities and public works, and reduce our myriad plastics and toxic synthetics back to benign, basic elements? Or are some so unnatural that they're indestructible?

And what of our finest creations—our architecture, our art, our many manifestations of spirit? Are any truly timeless, at least enough so to last until the sun expands and roasts our Earth to a cinder?

And even after *that*, might we have left some faint, enduring mark on the universe; some lasting glow, or echo, of Earthly humanity; some interplanetary sign that once we were here?[28]

So the question bears repeating: If not Nature/Society, then *what*?[29]

How will we feel the end of nature?[30]

"Nature is . . ."—what?[31]

. . .

Now that the stirrings of the earth have forced us to recognize that we have never been free of nonhuman constraints how are we to rethink those conceptions of history and agency?[32]

How can I even make this claim that forests think?[33]

Is it possible to understand our objects of study as inhabiting not only human time and culture, but also the nonhuman space and time of the planet, climate, or weather?[34]

How does the crisis of climate change appeal to our sense of human universals while challenging at the same time our capacity for historical understanding?[35]

Can temperature be interpreted?[36]

. . .

Is it too late to prevent climate change?[37]

Is it already too late?[38]

Once you realize how painfully small the size of our remaining carbon dioxide budget is, once you realize how fast it is disappearing, once you realize that basically nothing is being done about it and once you realize that almost no one is even aware of the fact that carbon dioxide budgets even exists [*sic*], then tell me what exactly do you do? And how do we do it without sounding alarmist?[39]

What responses are appropriate? By what methods, in what assemblies, and with what authority, will that appropriateness be determined, and in accord with which—or whose—interests, views, and priorities?[40]

How can we simultaneously be part of such a long history, have such an important influence, and yet be so late in realizing what has happened and so utterly impotent in our attempts to fix it?[41]

How did it come to this? How did any number of people end up embodying a contradiction that encourages them to shrug off the life-threatening consequences of events they foresee as impending, even as they make incremental changes to become more "environmentally conscious" and train themselves in new habits of worry?[42]

Do you see why I am worried? . . . Do you see why I am worried?[43]

Does this anxiety help us—perhaps by alerting us to the challenges we face? Or might it distract us or otherwise get in the way?[44]

But the question is—as so many have pointed out—whether it [the climate change movement] can attain that status and amass a social power larger than the enemy's *in the little time that is left*.[45]

Whether we frame this as a search for parables or for enemies, the underlying assumption is that meaningful action can be undertaken: it is not yet too late. But what if it is?[46]

Will it ever end, or will we start worrying about the problem after it is too late?[47]

So, what have we provoked?[48]

So my mind keeps coming back to the question: What is wrong with us? What is really preventing us from putting out the fire that is threatening to burn down our collective house?[49]

Then the thought came to me: What will future historians say about us? How will they answer this question?[50]

Is it too late . . . ?[51]

Who knows, for all the distance, but I am as good as looking at you now, for all you cannot see me?

What, now, do you have to lose? What else can you be but brave?[52]

. . .

So why are we not reducing our emissions? Why are they in fact still increasing? Are we knowingly causing a mass extinction? Are we evil?[53]

In an extinction event of our own making, what happens to us?[54]

What to *do* with a cracking, breaking, and leaking event, one that can neither be plotted effectively nor un-experienced?[55]

What I really would like to ask all of those who question our so-called "opinions" or think that we are extreme: Do you have a different budget for at least a reasonable chance of staying below the 1.5 degrees of warming limit? Is there another intergovernmental panel on climate change? Is there a secret Paris agreement that we don't know about? . . . And if anyone still has excuses not to listen, not to act, not to care, I ask you once again: Is there another Intergovernmental Panel on Climate Change? Is there a secret Paris agreement that we don't know about? One that does not include the aspect of equity? Do you have a different budget for at least a reasonable chance of staying below 1.5 degrees of global temperature rise?[56]

Of course, the devil is in the details—how to revolt? How to matter and not just want to matter?[57]

Fossil fuels should . . . be replaced by renewable sources of energy. But where, when, by what technical, political, and/or economic means, and through what social agencies? Are geo-engineered protections or mitigations (cement pilings, floodgates, cloud seeding, and so forth) altogether to be disdained, as suggested by those intent on purely socio-economic and/or spiritual transformations? If so, then when exactly should we give up driving cars and having children, and through what form of incentive and/or compulsion, administered and monitored by what governmental agencies?[58]

Of course, the devil is in the details—how to revolt?[59]

After all, how will thinking about Kant or Frantz Fanon help us trap carbon dioxide? Can arguments between object-oriented ontology and historical materialism protect honeybees from colony collapse disorder? Are ancient Greek philosophers, medieval poets, and contemporary metaphysicians going to save Bangladesh from being inundated by the Indian Ocean?[60]

How long will [they] keep us afloat?[61]

. . .

The problem has become for all of us in philosophy, science or literature, how do we tell such a story?[62]

How to tell this story so that it becomes more than elegy alone, both a record of these uncanny times and a rallying cry?[63]

How can we turn the long emergencies of slow violence into stories dramatic enough to rouse public sentiment and warrant political intervention, these emergencies that have given rise to some of the most critical challenges of our time?[64]

How do we find a way to tell these stories that keeps the heat on them?[65]

Now, how do we do that?[66]

How do we bring home—and bring emotionally to life—threats that take time to wreak their havoc, threats that never materialize in one spectacular, explosive, cinematic scene?[67]

"Where is home, and how do I get there?"[68]

As we head into stormier seas, we must ask ourselves, *If we cannot save the frozen Arctic, how can we hope to save the rest of the world?*[69]

How can we think in times of urgencies *without* the self-indulgent and self-fulfilling myths of apocalypse, when every fiber of our being is interlaced, even complicit, in the webs of processes that must somehow be engaged and repatterned?[70]

As a recent Wikipedia entry asks, with unintentionally absurd poignancy: "Why is oil so bad?" This query will be my refrain, and what I mean by it is why might twentieth-century petromodernity offer strong resistance to the imagination of alternatives, even as we recognize its unsustainability?[71]

Since we're imagining, why not also dream of a way for nature to prosper that doesn't depend on our demise? We are, after all, mammals ourselves. Every life-form adds to this vast pageant. With our passing, might some lost contribution of ours leave the planet a bit more impoverished? Is it possible that, instead of heaving a huge biological sigh of relief, the world without us would miss us?[72]

If we *will* not be missed, should we start missing ourselves *now*, in anticipation?[73]

It was objected to Hooke, that his doctrine of the extinction of species derogated from the wisdom and power of the omnipotent Creator; but he answered, that, as individuals die, there may be some termination to the duration of a species; and his opinions, he declared, were not repugnant to Holy Writ: for the Scriptures taught that our system was degenerating, and tending to its final dissolution; "and as, when that shall happen, all the species will be lost, why not some at one time and some at another?"[74]

Given, however, significant differences of situation as well as multiple, divergent interests among "us" (certainly among humans generally and even just among thoughtful, well-meaning academics and intellectuals), it is not clear, especially when equally justifiable interests conflict, which matters of concern should concern us most. Humanity? Every one of us? If not, then who? The biosphere? Every nook, niche, and creature? If not, then which ones? Future generations? How far into the future? If not thousands of years, then how many?[75]

But what happens when we are unsighted, when what extends before us—in the space and time that we most deeply inhabit—remains invisible?[76]

There is no food we can eat, clothing we can buy, or energy we can use without deepening our ties to complex webs of suffering. So, what happens if we start from there?[77]

Is this some community *we* rhizome into fragile connection to a place? Or a total *we* involved in the activity of the planet? Or an ideal *we* drawn in the swirls of a poetics?[78]

Poetics?[79]

Will you hold the end of the bundle while I braid? Hands joined by grass, can we bend our heads together and make a braid to honour the earth?[80]

. . .

Is there a word yet for the post-natural rain that falls when a cloud is rocket-seeded with silver iodine? Or an island newly revealed by the melting of sea ice in the North-West Passage? Or the glistening tidemarks left on coastlines by oil spills?[81]

What if we could make "anthropocene" a household word?[82]

Whenever a term or trend is on everyone's lips, I ask myself: "What other story could be told here? What other language is not being heard? Whose space is this, and who is *not* here?"[83]

What can human language, segmented into measured feet and clipped stanzas, signify against its "alternate song," the "endless cry" of "eternity"?[84]

Nowhere is this effort [to imagine new concepts] more urgent than in the effort to apprehend anthropogenic climate change: remember when *glacial* meant slow?[85]

Who needs an ice-breaker when you can count on melting ice?[86]

"Hey—did you see it, Will?! Did you see the ice-bird?"[87]

. . .

What if the [ecological] footprint measured, over time, on whom and what the nation's foot has trod—that is, who has paid for prosperity?[88]

Why should one include the poor of the world—whose carbon footprint is small anyway—by use of such all-inclusive terms as *species* or *mankind* when the blame for the current crisis should be squarely laid at the door of the rich nations in the first place and of the richer classes in the poorer ones?[89]

Who gets to see, and from where? When and how does such empowered seeing become normative? And what perspectives—not least those of the poor or women or the colonized—do hegemonic sight conventions of visuality obscure?[90]

I just thought I should ask: have you noticed?[91]

Is the image of nature as passive mud and dirt—a place where one leaves a footprint—really the best metaphor to capture the vitality of the web of life?[92]

Is it realistic to hope that those who are obsessed with maximizing profits will stop to reflect on the environmental damage which they will leave behind for future generations?[93]

How might capitalism look without assuming progress?[94]

Without progress, what is struggle?[95]

How to even comprehend this breadth of time? How to convey, despite our brevity as a species, the magnitude of our impact?[96]

What forces—imaginative, scientific, activist—can help extend the temporal horizons of our gaze not just retrospectively but prospectively as well?[97]

How can we come to know/think/feel/behave and subjectively experience ourselves—doing so for the first time in our human history *consciously* now—in *quite different terms*?[98]
Can we pull it off?[99]

. . .

And if the urgency of a subject were indeed a criterion of its seriousness, then, considering what climate change actually portends for the earth, it should surely follow that this would be the principal preoccupation of writers the world over—and this, I think, is very far from the case. But why?[100]
Why should this be?[101]
And why should I be studying for a future that soon may be no more, when no one is doing anything to save that future? And what is the point of learning facts when the most important facts clearly means nothing to our society?[102]
Is it already too late?[103]
Who gets to decide the answers to these questions?[104]
And if we disappeared, would—or could—we, or something equally complicated, happen again?[105]
This, then, is to be the way?[106]
Does it have to end this way? Does the last best hope for the world's most magnificent creatures—or, for that matter, its least magnificent ones—really lie in pools of liquid nitrogen? Having been alerted to the ways in which we're imperiling other species, can't we take action to protect them? Isn't the whole point of trying to peer into the future so that, seeing dangers ahead, we can change course to avoid them?[107]
So now what?[108]
So I ask / what else is there to hear?[109]

. . .

On Time: Fourth Experiment

FrankenClimate

> *Listen to my tale. . . . You will hear of powers and occurrences, such as you have been accustomed to believe impossible.*
> —MARY SHELLEY,
> *Frankenstein*

In the 1818 Preface to *Frankenstein*, Percy Shelley, ventriloquizing Mary's voice, recalls the now well-known scene of the novel's origins: several friends vacationing in Switzerland are forced by the inclement weather to spend an extended period of time inside. Today those friends might have resorted to their iPhones and other screens. Not so Shelley's friends. They decided to have a story-writing competition, the goal of which was to produce the supernatural affects that had impressed them in their reading together. They got to work. But then the weather changed, it became "serene" (6), as Percy puts it in the Preface, and, setting their pens and papers aside, they returned to their outdoor activities, their stories unfinished. Mary, however, remained writing, and her tale, the Preface concludes, was the "only one which had been completed" (6). This interest in completion and incompletion, the finished and the unfinished, spans the entire novel. Indeed, a few pages before the end of the novel, Frankenstein implores Walton "to undertake my unfinished work" and to kill the Creature (151).[1] Like Frankenstein in many of the earlier scenes, Walton prevaricates, and the Creature's fate is left unresolved. The novel itself has had such a long and vigorous afterlife, arguably, because its telling lends itself to "completion"—or to retelling—in so many different ways. It is as if Mary Shelley's novel itself is the "unfinished work" that refuses to die or be forgotten. Rather, it continues to live on in ever new forms.

One of those forms, I suggest here, is as a response to climate change. Climate change, while discussed in some contexts in the period, was by no means a vital public concern. While some critics have illustrated that *Frankenstein* does, in fact, directly engage with climate change (fearing not a warming climate but a cooling one) and many critics have considered its commentary on technology in light of what the novel suggests about the Anthropocene, I take a different approach.[2] I focus on *Frankenstein*'s deft command of its narrative frame and, in turn, its orientation to time. Can a better comprehension of framing offer a more robust response to the climate crisis? Can Shelley's novel not only be read as a commentary on progress narratives but also as an affirmation of post-time narratives? Does the way in which Shelley manipulates the novel's form contribute to different temporal modalities? On the one hand, *Frankenstein* can be read as a cautionary tale against progress narratives; it conveys a warning about the similar outcomes that may beset later readers if they continue to pursue territorial and technological progress. It is relatively straightforward, in this context, to read the monster as an allegory for climate change. On the other hand, that is not the whole story. For Shelley offers her account, of course, as a framed narration. By doing so, she invites us to notice the role of framing in any narrative account—but perhaps, especially, to notice frames in narrative accounts that have a monster, or a gaping hole, as their focus.

Throughout this book, I have been concerned to address the frames that make climate change intelligible (and unintelligible) to us today. I've been interested in how we often seek to straighten narratives and temporalities and extract lessons—indeed as I do, experimentally, in the *Middlemarch* chapter—in narratives that do not lend themselves to such straightening or lessons. In doing so, we often subdue the disorder that subtends any attempt to confront issues for which existing conceptual frameworks are inadequate. I close, anachronistically, with Shelley's *Frankenstein* because Shelley's novel so beautifully demonstrates what happens when we frame and hold open the frame at once. Or to put this another way: when both the frame and the gaping hole obtain and, following Agamben, possibilities for remaking time appear. To make this point, I exploit a kind of nested framing device in this chapter myself: the first three sections consider framing, interruption, and weather respectively; the fourth returns to interruption; and the fifth returns to framing.[3] In other words, this chapter seeks to replicate, very loosely, the framing strategy that Shelley deploys in her novel, albeit without the epistolary structure and different narrators. And where Shelley positions the Creature's narrative at the center of her text, I position the weather and a blank at the center of mine.

1. FRAMING (1): HEAR HIM

> "Hear me; let me reveal my tale."
> [Frankenstein to Walton]
> —MARY SHELLEY,
> *Frankenstein*

When I teach *Frankenstein*, the students tend to be most drawn to Frankenstein and his creation; they rarely remark on the frame narration, if they notice it at all.[4] Instead, they are intrigued by the Creature in relation to the outcast working classes, marginalized women, the perils attending new technologies, and so on. But the Creature's story is, of course, a story within a story within a story. The first frame is the letters that Walton writes to his sister, Margaret; the second frame is the story that Frankenstein tells to Walton (that Walton in turn relays to his sister); and the third frame is the story that the Creature recounts to Frankenstein (who in turn relays it to Walton who is relaying it to his sister). There are many interpretations of these nested frames and their impact on the novel overall.[5] For now I want to emphasize only the very basic point that these frames at once make us see how stories are always framed—mediated—in one way or another *and* to see how easily, in the telling, those frames are also obscured. The humanities themselves are particularly well equipped to address issues of framing, issues that, like the story of the scientist and his creation in this novel, tend to be read as if a neutral or nonframed account were available.

The outermost frame, the frame that contains all the others, comprises the letters that Walton sends to his sister. These letters call attention to "the interlocutionary relations" between listener and speaker, narratee and narrator. "Each act of narration in the novel," Peter Brooks writes, "implies a certain bond or contract: listen to me because . . ." (*Body Work*, 200).[6] It is widely agreed, indeed, that all three narratives "are clearly attempts at persuasion rather than simple accounts of facts" (Johnson, 3). This persuasion requires an interlocutor, a person who reads or hears one's story, and *Frankenstein*'s epistolary frame, as Gayatri Spivak writes, assigns Margaret as "the irreducible *recipient*-function of the letters (40). The reader of *Frankenstein* must, accordingly, interrupt or "*intercept* the recipient-function, read the letters *as* recipient" (40). Indeed, all narrative is motivated by the recipient-function and a framing function that *this* narrative, this novel, throws into relief.

These points also hold, of course, for my composition of this chapter. I seek to persuade you of my argument. And yet for this chapter to be true to Shelley's nested narrations, each of my sections would have to be told from a different point of view and in different contextual frames. What I want to draw out, instead, is the idea of moving toward a middle or center that finds its most powerful resonance not in its containment but in its dispersal through the other sections. I intercept the recipient-function as I read *Frankenstein*. What we forget, perhaps because it is so obvious, is that the recipient-function is also hot-wired for temporal disruption; it reverberates through time, backward and forward, and in doing so unsettles time in a way that also encourages us to occupy it differently.

Frankenstein's frame narration resonates with another frame narration focused, in a different way, on letters: Edgar Allan Poe's "The Purloined Letter." And Poe's short story, published in 1844, has generated its own series of responses that themselves operate as a kind of frame narration: the tripartite reading competition in which first Jacques Lacan offered an analysis of Poe's story, then Jacques Derrida, and then Barbara Johnson.[7] In his essay, Derrida writes: "The point is not to show that 'The Purloined Letter' functions within a frame . . . but that the structure of the effects of framing is such that no totalization of the bordering can ever occur. The frames are always enframed: and therefore enframed by a given piece of what they contain" (*Postcard*, 485).[8] This point speaks to Shelley's text in the context of its potent unfinishedness. But Shelley's text, in turn, speaks forward, anachronistically, to Derrida's text. It also speaks forward to me and invites me to think about climate change not in relation to a representation (the content of the frame, the many competing representations of climate change) but, rather, in relation to post-time and the effects of framing. In Poe's short story, the stable boundaries of Euclidean space are, Johnson argues, not "finite and homogenous" but dispersed and reversible (231). This disturbance of bounded space is also nicely demonstrated in Shelley's novel, but that is not where I want to go. Instead, I want to turn to a concept that these three essays on framing in Poe's story for the most part neglect: temporality and post-time.

If frames of reference disturb Euclidean spatial models, so too do they disturb temporal models of progress. If, as Johnson notes, "no analysis can intervene without transforming and repeating other elements in the sequence, which is not a stable sequence, but which nevertheless produces certain regular effects" (213–14), so, too, with every new frame, a new time is interposed. As the frames multiply, so too is temporal polyphony produced. To be sure, this temporal polyphony is minimized or eliminated in many well-worn strategies, not least of which is to erase the frame itself as "new" works circulate. Post-time, however, reintroduces this temporal polyphony: it upholds the sequence, upholds the palimpsestic temporalities that contradict it, and upholds the materiality of time that affirms and contradicts both. Written in 1845, Poe's short story does not conform to post-time in Eliot's narrow sense of accelerated temporality brought about by changes to postal delivery. But it reminds us of the link between postal delivery and time, the materiality of time, and concludes with its own (quoted and untranslated) letter to the future that has itself, in many ways, generated the critical commentary I have been discussing thus far.[9]

Eliot's narrator's narrow sense of post-time also does not obtain for Walton's letter-writing practice. Indeed, Walton's writing is exactly the opposite of post-time as Eliot defines it, reviving instead a kind of slow postal time that is reinforced in the novel's many references to the slowness and uncertainty of delivery. But what happens when we read the novel and its letters now? Judith Butler notes, following Benjamin, that the "reproducibility" of the image (in this case we might think of the republication of the novel) produces a "critical shifting"—what Derrida calls an "internal drifting" (*Postcard*, 489)—of contexts and frames (Butler, *Frames of War*, 9). This shifting and drifting happens because of the "shifting temporal dimension of the frame," a point that Butler makes as follows (in a passage that clearly conveys a debt to the Derrida passage cited above and is an instance of the temporal logic that it describes): "This very reproducibility entails a constant breaking from context, a constant delimitation of new context, which means that the 'frame' does not quite contain what it conveys, but breaks apart every time it seeks to give definitive organization to its content. In other words, the frame does not hold anything together in one place, but itself becomes a kind of perpetual breakage, subject to a temporal logic by which it moves from place to place" (*Frames of War*, 10).

The most troubling thing about Walton's letters home is not the geographical space that separates the two siblings—the "place to place"—but the temporal disturbances, the destabilizing "temporal logic," they produce. To be sure, neither Derrida nor Butler are focused on this aspect of temporality; they are instead focused on the temporal intervals by which the new contexts are implicitly defined. But postal delivery, as Eliot's narrator uncannily intuits, heightens this temporal logic by underscoring its complexity. Walton writes to his sister in his "now," hopeful that she will receive his news at a later date. As he writes he cannot know anything about his sister except what he knew when he left. She reads his letter, if it arrives, in her "now," unable to know what is currently happening in his life. That is, letters underscore with a particular sharpness, the palimpsestic quality of post-time; they invest in a sense of linear time even as they perform a palimpsestic temporality.

The novel, like the letters, is always arriving. It arrives as I read it and it arrives as you do. This should not be a surprise insofar as the twin origins of *Frankenstein* are ghosts stories and dreams, those genres that invoke the liminal spaces of life and death, waking and sleeping. But it underscores the last framing device I want to discuss here, a frame that all the interpretations of *Frankenstein* share but to which we do not often refer: the frame of readers as themselves always already enframed. That is, we can discuss Margaret or ourselves as serving the recipient-function of these letters but our reading, too, always comes with its own frames indefinitely multiplied and, as Butler notes, "in a kind of perpetual breakage." There are the frames of Shelley's early readers (by no means singular or predictable) and there are the frames of readers extending indefinitely into the future. In my case, I develop the frame of climate change not only as an arena of academic inquiry but also, more than many of my other academic areas of study, as an arena of personal, social, political, and existential concern. And uniquely, a concern that does not give me a firm place to stand: I feel daily the strangeness of pursuing the academic activities I enjoy while also feeling, simultaneously, that our daily lives have been disturbed, in ways for which we have not yet fully accounted.[10]

2. INTERRUPTION: ARRIVING, WAITING

> *How slowly time passes here.*
> —MARY SHELLEY,
> *Frankenstein*

Frankenstein is a historical novel.[11] Written in 1818, it returns to the 1790s, the same period, indeed, to which *Adam Bede* also returns. Shelley's novel reflects on the period of her own origins, whereas Eliot's novel reflects on generational change. These are not the temporalities of deep-time, then, but of human lives. And yet the 1790s mark a transition that has a bearing not only on the comprehension of nature and human life but also on time itself.[12] This period introduces "the threading of knowledge with time" (274), as T. H. Ford puts it, paraphrasing Foucault. The new comprehension of knowledge as intertwined with time would have been well established twenty years later when Shelley was writing. I've noted that the letters Walton writes his sister straddle both their time and ours in a state of always-arriving, and, in many ways, the entire novel is written in the mode of anticipatory arrival that those early letters introduce. Let's look more closely at how the letters function in the novel.

"I arrived here yesterday," Walton writes his sister Margaret (his letters are all addressed to *Mrs. Saville, England*). "Here" is St. Petersburgh,[13] Russia, a way-station on his "voyage of discovery" to the North polar region (14). Nineteenth-century readers easily absorbed the fractured temporality of the letter—written then, read now—as do we. The datedness of letters is part of the form; the time of writing (in this case, Dec 11th, 17--) is not the time of the recipient's reading. As noted above, this point is even more pronounced for readers of the novel. We seek to occupy the position of the addressee (the first word of the novel is "You" [7]) but, at the same time, we read the words much later than their period of composition—not just weeks or months later but sometimes years or centuries later. Letters accordingly introduce reflections on time and post-time that push back against progress, bound up as they are with relay, delay, and postal services.

Letters help us to navigate discordant temporalities. Their material form reminds us of the layering of time: the past that a letter animates in the present-reading as well as future returns and rereadings. They embody the carrying forward of time. Walton's letters are also *about* time: slowness, delay, foretaste, and interruption. They confirm that Margaret's "evil forebodings"—her premonition of future events—have not come to pass; Walton's arrival (for now) dispels her concerns. In sharp contrast to Margaret's "foreboding," the "cold northern breeze" Walton feels as he walks the streets of St Petersburgh, is a "foretaste of those icy climes," inspiring "delight" (7). Margaret's foreboding and Walton's foretaste look to an unknown future in conflicting registers. That future is also imagined by Walton in a "day dream" (itself overlaying the night on the day) as following different temporal coordinates in which the day is erased all together: at the pole "the sun is for ever visible" producing "perpetual splendor" and "eternal light" (7). Walton imagines harnessing a knowledge that would give him command of time and space, the magnetic needle that guides his travel, and the forces that "regulate a thousand celestial observations" (7).[14] He may, he writes Margaret, "tread a land never before imprinted by the foot of man," likening this joyful journey to "an expedition of discovery" up a river "in a little boat" (8). Most of all, however, he might find "a passage near the pole" through which to travel more speedily and conveniently to distant countries (8).[15] His passionate writing about the possibility of discovery, then, is not just threaded through time but a particular version of time invested in progress that the novel, overall, unsettles.[16]

This opening section comprises four letters, the last one of which introduces Frankenstein and reads as a series of diary entries.[17] The first letter is written in December from St Petersburgh (in the period when Mary Godwin became pregnant with Shelley). The second letter is written in March from Archangel (on the day before Godwin and Wollstonecraft marry).[18] The third and fourth letters, dated July and August respectively, are both written en route from the ship Walton has secured for his expedition of discovery. As if to reinforce the baffling of temporality, the second letter begins: "How slowly time passes here [in Archangel]" (9). Indeed, the second letter marks a period of waiting for bad weather to pass and Walton's journey to begin: "My voyage is now only delayed," he writes, "until the weather shall permit my embarkation" (11). The third letter, by contrast, written over three months later, begins in a spirit of speedy dispatch: "I write a few lines in haste" (12). Walton's travels are underway, the winds bode well, and he has nothing eventful to report.

The fourth letter, written exactly a month later, returns to weather-incurred delay; Walton's ship is "nearly surrounded by ice" and "compassed round by a very thick fog" (12, 13). It cannot move. And then, in a scene of endless white, the Creature's sledge emerges and, as many critics have noted, moves across the snow like a line across a page before he disappears (off the page, as it were). Hours later, out of the white, Frankenstein arrives by the side of the boat, cold, sick, depressed, and obsessed. Walton's period of waiting is interrupted by Frankenstein's arrival. The record of his arrival—introduced as a strange "accident" (12)—and growing friendship with Walton is communicated at "intervals" (15) across three diary entries roughly one week apart from each other. The content of these letters roams widely through temporal periods: they envision a future of discovery; recall a childhood and young adulthood marred by a neglected education and a failed effort to become a poet; and record the present with its weather, interruptions, and accidents. The temporal gaps between these letters is effaced in their novelistic presentation as is the letters' journey along the post road and what would have been dramatic seasonal changes from December through July.

While Walton's journey is just beginning, Frankenstein's "fate is nearly fulfilled" (17). Responsive to the possibility of interruption that another interlocuter brings, Frankenstein tells Walton (and by indirection Margaret and us) that his is a story of "fate" and "destiny" in which the outcome is "irrevocably . . . determined" (17). He asks only that Walton "listen" when "he should be at leisure" the following day (17). Walton tells his sister that he has "resolved every night, when I am not engaged, to record, as nearly as possible in his own words, what he has related during the day. If I should be engaged, I will at least make notes" (17). Just as we often forget the epistolary frame, it is easy to forget that the story of Frankenstein and the Creature is a reconstructed narrative. Frankenstein tells his story during the day and Walton listens; later, "every night," Walton either records what he remembers of the day's story or writes down notes for further elaboration at a later date. The composition of Frankenstein's story, then, like Walton's day dreams, weaves together night and day as it also weaves together the voice of Frankenstein (itself multivocal) and the voice of Walton.

The four opening letters close in the 1818 edition with Walton's enthusiastic declaration that "in some future day!" he looks forward to reading the account he sends Margaret as if this collaborative account is a guarantee of that future (17). The letters are meant not only to travel to Margaret but also, after an extended temporal gap, to arrive back in Walton's hands for future perusal. But the revised 1831 edition modifies—and in some senses forecloses—that future. It appends an additional description of Frankenstein and ends instead with the following words: "Strange and harrowing must be his story, frightful the storm which embraced the gallant vessel on its course and wrecked it—thus!" (38). His account will not be, then, a pleasant story of going up a river in a little boat.

What moves me most in these four letters that open Shelley's novel is the first line of Letter IV (the letter that comprises the remainder of the novel): "So strange an accident has happened to us, that I cannot forbear recording it" (12). *I cannot forbear recording it.* I cannot *not* write this story, Walton in effect tells his sister. Suspended between Margaret's "evil foreboding" and his own delightful "foretaste" of things to come—suspended, one might say, in the gaping hole—Walton sits down to write. Far from shore, travelling in a boat across uncharted waters, Walton records a story that itself moves at once forward and backward, that is made of recollections and fragments stitched together like the Creature himself, that is about endurance, duration, accident, disaster, and catastrophe.[19] He presses onward, fueled by the "prospect of arriving" (8) but also haunted by his past failures.

Frankenstein is motivated to tell Walton his story, despite his weakened state, because he wants it to be a message and a warning to Walton (and others) to avoid what he has done. Walton sits by Frankenstein's bedside, attentive to this man who resembles "a celestial spirit, that has a halo around him" (16). This warning will necessarily pass through Archangel, travelling the post road back to St Petersburgh, and on to England. I imagine the newly named Archangel, the first angel but also the new angel (or, fancifully, the *angelus novus* with its own cancelled-out warning), situated on the shore of the Baltic Sea, as a kind of switch-point or relay allowing

Frankenstein's letters to reach not only Margaret but also me on "some future day!"

3. BAD WEATHER, BLANK

> *The play bill amused me extremely for in the list of the dramatis personae came, — by Mr. T[homas] Cooke: this nameless mode of naming the un[n]ameable is rather good.*
>
> —MARY SHELLEY
> (letter to Leigh Hunt)

Benjamin's *angelus novus* is poised in the midst of the storm of progress; "wreckage" rises at the angel's feet as he stares fixedly at the past. Shelley's Archangel is the place where Walton waits out the storm before departing on his expedition of discovery.[20] Bad weather forces him to delay his departure from Archangel, it forces him to stop when, underway, he is surrounded by ice and mist (and Frankenstein gains access to Walton's ship in this moment of forced arrest, his own body's vessel "wrecked" in the figurative "storm" he relates), and it forces him to stop at the end of the novel, as Frankenstein sickens and the Creature again emerges out of the mist realizing the book's arc and Frankenstein's, by this point, fading hopes.

But for a week beginning on August 19, 17—, Walton's progress toward the Northeast Passage and the weather he endures (or not) is eclipsed by Frankenstein's story (the week roughly right before Mary's birth). The narrative does not return to Walton's account until August 26, and we are not informed of inclement weather again on the voyage until September 2. In place of Walton's weather reports is Frankenstein's narrative in which he aligns bad weather with the emergence of the Creature, and at the heart of his narrative—the heart of the storm, so to speak—is the Creature's testimony. The middle of this narrative, then, not only introduces the strange "accident" of Frankenstein's arrival, but also a sense of possibility, ignited by listening to the tale, that disrupts the propulsive progress of the story of Walton's expedition (about which we hear nothing during Frankenstein's story).

Derrida's grievance with Lacan's reading of "The Purloined Letter," is that he lifts out—extracts—the center of the text and fails to see that as a result he does an injustice to its meaning. As Derrida puts it, "He drop[s] the frame" (*Postcard*, 431). As I noted above, this is also what readers often do with *Frankenstein*. That the story of the monster is the content of a letter home is a significant dimension of the story. But there is also a "blank" at the heart of Poe's story that both Derrida and Johnson caution against filling in too recklessly. Indeed, filling in the blank suggests that it is a container rather than a necessary condition for meaning. I thought about this point as I cavalierly entertained the idea of filling in the monster's figurative blank as weather. As so many critics have noted, the monster lends itself to an infinite regress of allegorical associations as it also lends itself to a dizzying slippage, doubling, and mirroring of the other characters. The monster, we might say, gives form to all these things—the gaping hole, infinite regress, doubling, and mirroring—that we cannot say. The monster defies categories and demonstrates the havoc such category defiance produces.

Staying with this point would be to stay with the blank, to resist filling it in. This is exactly what Shelley recommends when, after seeing the first public performance of her novel as a play, she expresses her appreciation for the playbill's reluctance to name the Creature, noting his name only as a blank, an "——." "This nameless mode of naming the un[n]ameable is rather good," she writes to Leigh Hunt in 1823. The word "blank" does not appear in *Frankenstein*, but interestingly Shelley refers to it twice in her 1831 Preface to the novel. Recalling her fondness for fantasy and daydreaming, she describes a period of time she spent in Scotland as an early teenager; while there, she writes, her "habitual residence was on 'the blank and dreary northern shores' of the river close to the house where she was staying (169). "Blank and dreary on retrospection," she corrects, but "not so" at the time (169). At the time, Shelley filled the blank in, so to speak, with her fantasies, fantasies that, like the play-bill, serve as a "nameless mode of naming the unnameable."

With her friends in Switzerland, she confronts another blank that is again crossed over. She keeps trying "*to think of a story*," specifically a ghost story, to "rival" established stories but instead feels only that "blank incapability of invention which is the greatest misery of authorship" (171). This blankness continues for several days. And then Mary has a dream. In the dream she sees the now-familiar "pale student of the unhallowed arts," "his odious handy-work," and "the hideous phantom." She wakes up terrified, and it is this terror that fills in the blank, so to speak; she announces the following morning, "I had *thought of a story*" (172). Slipped between the "blank" and the gripping outline of the story, however, is an account of temporal complexity to rival this story of her novel's origins. "Every thing must have a beginning, to speak in Sanchean phrase," she writes, "and that beginning must be linked to something that went before" (171). This is the point that Freud would also make about dreams themselves. Shelley makes a "transcript" of what she calls her "waking dream" beginning with the sentence, "It was on a dreary night of November" (172), the now famous opening to the Creature's emergence. She fills in her blank with a monster.[21] Which is another way, as she notes in her comment to Hunt, of leaving the blank blank, of staying with the terror and making others feel it too. To return to Shelley's comment on beginnings, it is also the blank, or the ". . .," of beginnings. The blank blank of the monster is what stands in for unnameable relations, for all that we cannot stabilize with a name or a category, however much our terror would be eased by doing so. Shelley's account of the dream is an origin story, but, as she well knows, the origin is not an origin but only another story—in this case, a waking dream—that recedes indefinitely. This point is underscored by the double blank of this origin story itself, first recurring to the blank desolation of a Scottish shore and then in the blank despair when inspiration fails.

I am not suggesting that the monster is about the impossibility of meaning. Indeed, I am suggesting the opposite: that the monster touches the very nerve of meaning, and that its charge often makes us jump back or recoil. The monster touches that place that reminds us that all of our meanings are inventions, things we make up, and that while some are sturdier than others, none can be affixed to an origin that prevents their waver or wobble. I don't read the monster as a warning against invention but rather as a warning against not attending to this dimension of monstrousness that Shelley also calls a "haunt[ing]" (172).

Anthropogenic climate change may be a product of the same heedlessness to the implications of invention that Frankenstein displays. But it will never be adequately addressed using Frankenstein's principles. Indeed, he exhibits, in one person, the default responses of those in the Global North—a gasp of horror, disbelief, looking away, fleeing, blaming others, disavowing or displacing responsibility—all the while being fully imbricated in and defined by what he has created. An adequate response to the climate crisis needs something more like Shelley's willingness to look into the blank blank, to open it for others, to animate what cannot be named, and to stay there despite the discomfort, fear, and sense of dislocation it produces. Shelley's own stab at the unnameable is the monster's constantly shifting designations that recall, for me, Benjamin's proliferation of different terms for interruption. With respect to the monster and interruption, the relational tension that the different appellations generate, is especially pronounced insofar as the words sought gesture toward that unnameable relation that I've been calling the gaping hole.

I have digressed from my intention of filling in the blank with weather. What happens when weather is the center of the story? To be sure, weather, even when it goes entirely unnamed, is always there. And, as Heidegger reminds us with respect to things, we tend to notice it only in its extremes: bad weather and brilliant weather. But while bad weather is a trope of gothic fiction, it would be a stretch to suggest that it is in any way the center of this story. Indeed, it would be more accurate to suggest that it is its impetus. Just as bad weather creates the conditions for Shelley's story (storytelling inside with friends), so too is it the spur both to Walton's story (it affords him the time to write his long letters to Margaret), and to Frankenstein's interest in chemistry, technological discovery, and subsequent "invention"—the word that Shelley also uses for storytelling—of the Creature. But bad weather also presages the Creature's many appearances in the novel, in turn encouraging readers to align the Creature with bad weather accordingly.

Just as Frankenstein implores Walton to "listen to his history" and learn from it, so too Waldman implores Frankenstein to listen to his, and to avoid trying "to become greater than his nature will allow (31). Frankenstein does not listen. He creates the Creature as "the rain pattered dismally against the panes" and the rain continues through a night of "wildest dreams" (34). Frankenstein wakes to see the Creature's eyes "fixed on" him; he flees into the "dismal and wet" morning, hurrying on as the rain "poured from a black and comfortless sky" (35). The Creature's birth coincides with bad weather, and every encounter he has with Walton thereafter is also marked by bad weather. The next time Frankenstein sees the Creature (shortly after the death of William), for example, a storm rages and "thunder burst with a terrific crash" over his head (48). Lightning illuminates the surrounding country and the lake appears "like a vast sheet of fire" (48). The Creature emerges in this dramatic light show, illuminated by the flashes. They don't speak. Frankenstein's second encounter with the Creature is also ushered in by a storm when he is on a visit to Chamounix. The weather had been fine but soon clouds over, and "the rain poured down in torrents" (63). Frankenstein, alone on the glacier, describes the rain that "poured from the dark sky" and the "sea of ice" he can see below (64). The clouds clear, a breeze blows in, and the uneven surface of the glacier rises "like the waves of a troubled sea" (65). At this point he sees the Creature coming toward him, and this time, instead of retreating, Walton confronts him.

The Creature convinces Walton to listen to his story, and for the duration of the Creature's account, the bad weather recedes. To be sure, it sometimes rains and the Creature is often cold, but the stormy weather that punctuates Frankenstein's account is missing in the Creature's. Bad weather returns, however, on the next occasion that Frankenstein sees the Creature. Indeed, by this point in the novel, as soon as thunder and lightning are introduced, we expect the Creature to soon follow. The wind "rose with great violence," "clouds swept across the sky," "restless waves" disturbed the lake's surface, and "suddenly a heavy storm of rain descended" (135). This is Frankenstein's long-awaited and feared wedding night. And, sure enough, the Creature soon appears. He kills Elizabeth and, now merged with the storm itself, narrowly eludes Frankenstein's pistol shot as he, "with the swiftness of lightning, plunged into the lake" while "rain [again] fell in torrents" (136–37).

Thus far I have not addressed the Creature's story but only Frankenstein's story of it, an account that, to follow my plan in this chapter, should have been the focus of the "interruption" frame that precedes and follows this one. On the one hand, the frames do not contain the story; the Creature is introduced, after all, in Walton's narrative. On the other, what is most relevant about the Creature, for my purposes, is its blank. If the monster is the gaping hole or interruption at the heart of this story, we should listen to his words. For he, too, tells a story of persuasion. When we listen, as so many critics have noted, we hear an account not of something cruel and uncaring but of something maimed and hurt, wounded at its heart, and of something almost magical: the desire to heal that wound and find relation. What the monster seeks is both a possibility of relation and a renewal of species. It is the yin and yang of the monster: utter devastation or species renewal. Both are possible. But to navigate that territory, one has to attend to the story, listen closely, stare into the gaping hole, and feel the blank blank of possibility that trembles there.

And if the monster is an allegory for climate change? If he inflicts harm "upon everlasting generations" (114) and portends, as Frankenstein fears, the destruction of "the whole human race"? What then? There are two ways that climate change tends to get addressed in this novel: the first typically takes its point of departure from the eruption of Mount Tambora in Indonesia in 1815 and the weather disturbances this eruption generated over the next three years across Europe and elsewhere. It provoked, Gillen D'Arcy Wood argues in his groundbreaking book on this topic, widespread droughts, economic depression, and extreme weather.[22] It also generated discussions of climate change and global cooling in relation to Arctic exploration.[23]

The second climate-change-inflected approach to *Frankenstein* takes up the novel as a document of the Anthropocene. In Walton's first letter to Margaret, he rhapsodizes that he will "tread a land never before imprinted by the foot of man" (7). It is precisely the imprint of the foot of man on remote lands (as well as close ones) that, for many, describes the Anthropocene: the carbon footprint serves as a metaphorical shorthand for the now measurable human impact on earth systems. Walton's rationale for his dangerous voyage is the narrative of progress that also underpins Anthropocene thinking: "You cannot contest," he writes Margaret, "the inestimable benefit which I shall confer on all mankind to the last generation" (8). And toward the end of Frankenstein's account, he, too, recalls the imprint of the human (or human-made) on the natural environment when he describes, with dismay, the "print of his [the Creature's] huge step on the white plain" (141). From its earliest critics on, *Frankenstein* has been read as a warning against the unchecked development of new technologies. Sir Walter Scott wrote in the *Edinburgh Magazine*, for example, that Shelley was illustrating "that the powers of man have been wisely limited and that misery would follow their extension" (195). And Wood more recently notes that the novel tells a "cautionary tale" that warns "against the technological hubris of our modernity through figures of intense and widespread suffering" and imagines "our own climate 'Frankenstein,' a Creature who feeds on carbon waste and grows more violent by the year" (234).

We may now indeed be close to that last generation to which Walton refers. And our Anthropocene ambitions, like Walton's and Frankenstein's—figuratively written as monstrous footprints in the snow—have multiplied those footprints across the white plain to the point where the white plain may soon be gone entirely.

4. INTERRUPTION: RETURNING; OR THE GHOST OF FRANKENSTEIN

> *"We talk of ghosts."*
> —MARY SHELLEY,
> journal entry

In his first letter Walton muses: "And when shall I return?" (9). If he succeeds, he tells Margaret, it may be many years. And "If I fail, you will see me again soon, or never" (9). In the closing letters of the novel, "never" is looking most likely. The reader "returns" to the frame of the letters at a break or interruption in the text marked "WALTON, *in continuation*" (145). There is a break *and* a continuation. The text then doubles back to those opening letters and the timeframe of the letters' composition. Immediately before this break, Frankenstein describes how he pursued the Creature until the work of the weather "extinguished" his hopes of revenge (144). "The wind arose," he says, "the sea roared; and, as with the mighty shock of an earthquake, it split, and cracked with a tremendous and overwhelming sound. The work was soon finished: in a few minutes a tumultuous sea rolled between me and my enemy, and I was left drifting on a scattered piece of ice, that was continually lessening, and thus preparing for a hideous death" (145). His life is extended only by the fortunate circumstance of Walton's ship, but if the work of the storm "was soon finished"—a work that notably was about breaking the foundation, as it were, on which he stood—his own "task is unfulfilled" (145). Frankenstein's account, moreover, closes poised not only between the "work was soon finished" and the "task is unfulfilled" but also between life and death.

In his final words to Walton, Frankenstein imagines his afterlife as a ghost, suspended in the air, ensuring the end (but what end?) that he seeks: "I will hover near," he says, "and direct the steel aright" (145).

Indeed, Frankenstein's ghost continues to "hover near," transposing one ghost story onto another. We pass from the steel of Walton's sword to the interruption—"WALTON, *in continuation*" (145)—and Walton again at his desk writing his sister, recording what the seemingly smooth form of Frankenstein's account, unlike Walton's, has not conveyed: Frankenstein's broken language, agitation, bodily distress, and rage. "Sometimes, seized with sudden agony, he could not continue his tale; at others, his voice broken, yet piercing, uttered with difficulty the words so replete with agony" (145). Sometimes his voice was "tranquil" and sometimes it was "like a volcano bursting forth" (146). The tale, Walton tells us, is begun, interrupted, begun again, just as the frame narrations also break, and break again, the narrative's overall sequence.[24] There were times when Frankenstein "could not continue his tale," but in this "continuation" Walton reminds us of how those breaks are effaced, how the gaping chasm between two ice floes can be made to seem nothing at all. He also reminds us that we hear Frankenstein's and the Creature's stories, and we hear them not. In other words, to read this novel as a simple allegory of climate change goes against the grain of the dialectical image it ignites; it translates this story into a tale with a message that we can carry into our future as a lesson. But that's not what the novel asks of its reader.

What does it mean to return? Walton's ship is "surrounded by mountains of ice" making progress impossible; fearing that they will not be released by the ice into a "free passage," the sailors implore Walton to turn around. But Frankenstein, in his weakened state, intervenes. His efforts may be for naught, but he has by no means relinquished his ambition or his hopes for human intervention in the nonhuman. He tells the sailors they should embrace their potential role as "benefactors" of their "species" and should risk death for the "benefit of mankind" (149), words that mirror Walton's opening comment to Margaret about his desire to benefit "mankind . . . to the last generation." Walton also echoes Frankenstein's views on returning: "How all this will terminate, I know not; but I had rather die," he writes Margaret, "than return shamefully,—my purpose unfulfilled" (150). Instead of the open question (When shall I return?) from earlier, then, he now cleaves to his ambition in a passage that could serve as an uncanny motto for the Anthropocene.

The reader is surprised, in this context, to read the following letter, the shortest of the series of five letters that close the novel: "The die is cast; I have consented to return" (150).[25] What happened? Walton listened to his crew and respected their appeals to turn around; he is not reconciled to this decision, but he respects it. This is a powerful commentary on community, however muted by the narrative passing over the conversations and debates that led to its outcome. *Frankenstein* is often applauded for its cultivation of sympathy and its commitment to forging connections (captured, not least, in Walton's letters to Margaret). But this moment when Walton *listens* to his crew and follows their communal wish *despite his ambition* offers an extension of this point pertinent to our own moment.[26] How can we harness interruption now? How can we learn to listen to the "tales" that circulate around us?

"I am interrupted" (152), Walton abruptly writes. Turning around—returning—also enables the completion of Walton's tale through another conversation that *is* recorded in the text. Walton is in the process of recording Frankenstein's final words in a letter to Margaret as he journeys back to England. Here there is a pause in the text. We only know that between one paragraph and the next Walton has stood up to investigate the unusual sounds that have interrupted his writing. It unfolds, however, that what occurs in the interval is what makes the completion of his story possible. It is not his arrival at the pole and it is not the killing of the Creature. The "tale which I have recorded would be incomplete," Walton writes Margaret, "without this final and wonderful catastrophe" (152). Walton refers, of course, to the scene of the Creature at the bedside of the dead Frankenstein.

When the Creature sees Walton, he springs toward the window to leave, but Walton, importantly, calls him "to stay" (152). The Creature "pause[s]" (153). He looks at Walton and then looks away, returning his focus to Frankenstein. In this moment, thoughts of finishing Frankenstein's "unfinished work" and killing the Creature are "suspended" (153). Frankenstein's ghost may be hovering near, but it does not direct Walton's action. Instead, "in a pause" (153) of the Creature's apparent rage, Walton broaches a conversation in which he reprimands the Creature, saying that his "repentance" is, in effect, too late. The Creature then schools Walton in the complexity of human actions, emotions, and remorse. One can do things one regrets.

Initially moved by the Creature's story, Walton is brought short by his recollection of Frankenstein's warnings about his eloquence. He counters: "You throw a torch on a pile of buildings, and when they are consumed you sit among the ruins, and lament the fall. Hypocritical fiend!" (154). But the Creature "interrupt[s]" him again (154) and renews the passionate speech outlining his own unhappiness and reminding Walton not of his singular responsibility but of the collective actions that have brought them to the moment they share together: "Am I to be thought the only criminal, when all human kind sinned against me?" (155).[27] The Creature then returns to the question of unfinished work that has haunted this novel as a whole by noting that his "unfinished work" coincides exactly with Frankenstein's and, by extension, Walton's. All that remains to be done, the Creature says, is to kill himself. And yet, of course, the novel famously ends on a note of suspension with the Creature disappearing in "darkness and distance" (156).

In my earlier discussion of "fore"—foreboding, foretaste, forbear—one fore word to which I did not refer was "foreshadow." *Frankenstein* is full of foreshadowing, indicators in the earlier text of later things to come. As Nathan Hensley reminds us, foreshadowing itself occupies an impossible place: we cannot know that something foreshadows something else until the latter event occurs ("Database"). At that point, we *return* to the earlier event (or text) and reread it as foreshadowing—that is, intimating things to come. It is always only foreshadowing after the fact. These comments also relate to post-time's overlapping temporalities as well as its gesture to after-time and futurity. While there is much foreshadowing and anticipation of future events in *Frankenstein*, is it possible also to discuss novels as foreshadowing, or anticipating, things of which they could not possibly be aware? Does *Frankenstein* foreshadow climate change and species extinction, for example? Walton refers to Frankenstein's death as an "untimely extinction" (152), a phrase that anticipates the Creature's call for his own "extinction" in the closing paragraphs of the novel (156); these intimations of extinction arguably resonate in new ways for us now. They anticipate a future that has not yet arrived and that we only read retrospectively and recursively. Foreshadowing also brings me back to—returns me to—my comments on temporality and not having a place to stand that I noted above with respect to climate change. Here I take a cue from Frankenstein's parting words and invoke his ghost to hover here, conjoined with Mary's, and to turn the steel tip of a sword into the steel tip of a pen that continues the story.

5. FRAMING: HEAR HIM NOT

> *"You have read this strange and terrific story."*
> —MARY SHELLEY,
> *Frankenstein*

We are still reading this strange and terrific story, hearing it and hearing it not. The outer frame is the letters that Walton sends to his sister, missives binding his connection to a community beyond his voyage and the relationships he forges there. In the conceit of the epistolary novel, they are also missives sent to us, its readers. We return to the present time of Walton's writing with the line to Margaret, noted above, "I am interrupted." Walton is interrupted by a sound the Creature makes, presumably as he gains access to the boat through a window. Just as Walton's early letters record delays in Walton's journey and the interruption to his travels following from bad weather, so too we return to the boat stalled in bad weather. And just as the main event of the earlier letters is Frankenstein's unexpected arrival on the boat, so too is the main event in this closing frame of the novel the Creature's unexpected arrival on the boat. Walton's voice here gives way to the Creature's, and it is the Creature's words that not only comprise the center frame of the novel but also spill out to fill the narrative's closing pages, appended only by a brief, final paragraph from Walton's perspective.

Indeed, all of the novel's frames falter in the narrative's conclusion. Not only are the central characters each other's doubles, as many critics have noted, but also the frames that define them extend outward toward a limit that is ever receding, as they also fold endlessly in toward the center they define. But they perhaps falter most in that interruption that is the Creature's: he breaches the boat and enters Frankenstein's cabin. On the one hand, the Creature does not persuade Frankenstein to make a female companion for him, and Frankenstein does not persuade Walton either not to listen to the Creature or to kill him. He listens and the Creature lives. On the other hand, does Walton's account, as told to his sister, Margaret Walton Saville, persuade his sister and by extension us, to listen and perhaps to change our minds? If so, what might the monster's interruption to the story have to do with it? In addition to refusing the sequential stories that the Creature and Frankenstein individually relate, this novel overlaps and braids multiple temporalities together. Margaret as recipient-function, to recall Spivak's term, underscores this point: the letters are sent to a recipient whose initials clearly invoke Shelley's dead mother, Mary Wollstonecraft Shelley, and they are sent to us (Wolfson). They ask us to imagine the ghost of Mary's mother sending these letters to our future, inviting us to use them in a post-time we are only beginning to understand.

"I am interrupted," Walton writes Margaret. This is a description of an event. But it can also be read as a description of Walton. I am a being who is interrupted. I am not self-contained. I am broken and interrupted. I am a ghost writing with a ghost's pen. I write as the ghost of Frankenstein guides my pen, as the monster's "spirit," too, shapes my words.[28] In this last section, it is the Creature that interrupts Walton's writing, insisting on his importance to the story that is being imparted. He is necessary, as I noted above, to the completion of the story. Walton needs the gaping hole, the place where the frame wobbles, the place where definitions give way to uncertainty, where progress narratives transmute into post-time narratives. The story's completion is its incompletion, its breathing spirit, its ghost hovering near. There is a monster at the heart of every story reminding us that it could be otherwise, that this is only one way of telling the story, that the story is precarious, provisional, and indelibly bound up with its framing, its mediation. Sometimes these monsters can barely be heard and sometimes they roam widely, visible on every plane and in every line. We also need a monster at the end of every story reminding us that it is not over yet, and that the possibilities for which the monster is a placeholder remain.

"I am interrupted," Walton writes. And I interrupt myself with these nested sections. What does it mean to be interrupted? What possibilities, if any, does it produce? In our digital age, interruptions often are aligned with the distractions that prevent sustained thinking. But, as I have argued in this book, there is also a long tradition of interruption serving philosophical thinking by introducing a pause. This pause provides a space for thought as well as an attentiveness at once to framing, mediation, and the instability of the palimpsestic, multitemporal, discordant temporalities I have been calling post-time. Shelley's novel dislodges sequential narrations to privilege instead multiple temporalities, connections, and conversations, as it also, anticipating Benjamin and co-writing across the centuries, realizes interruption as a temporal dissolve that manifests other possibilities, and that preserves and cancels at once.

In the end, instead of killing the Creature, Walton listens to him. He forges a connection. This strikes me as the sort of response to climate change that humanities approaches enable: connection, conversation, and a willingness to consider the unexpected as opposed to vanquishing it. At the same time, this willingness to introduce a conversation that rewrites the line of linear progress, so to speak, also trains us in the temporalities of post-time. The weather is intimately bound together with everything else—human and nonhuman. Like Walton, there is an option to *turn around*, redirect our course, consider what survival requires. Like Walton, we have an option to acknowledge our shared intimacy with weather, however uncomfortable. Turning around does not mean disavowing what has been done; it does not mean going backwards. It means rethinking the backwards and forwards, the before and after, the models of temporality that privilege only linearity. It means facing in a different direction, which could also be multiple directions; it means staying with the gaping hole. It means changing one's course and doing things "quite otherwise."

I return, then, to the frame of my own reading. The idea of writing a nested chapter was an experiment. As with the other experiments in this book, I am not certain of its success. *But* in this period of climate crisis, I find that it is important to experiment even if it means courting failure. For a topic that defies representation and for which the accumulation of information seems only to dull responses, experimental approaches can perhaps have some traction where previous interventions have not been fully effective. I was struck, for example, by Anne Enright's comment that *Frankenstein* "would never have survived a creative writing workshop" and yet its very "bocketty"-ness signals its success (9). I was also struck by how Walter Benjamin, when he was grappling with ideas that were most resistant to prevailing representational modes, sought experimental forms. *Frankenstein* is not just a commentary about the consequences of altering the natural world, of producing the world we now call the Anthropocene. It is also a commentary on an alternative. This alternative emerges at what we might call the formal level of the novel: its pauses, interruptions, framing, and unfinished inquiries. And the novel's own "unfinished work" animates the vitality of its conversations and collectivities and enables it to reach across the centuries to participate in our ongoing climate conversations today.

The Creature's leap out the window on to his ice-raft also returns me to the bridge over the River Tay with which I opened and the gaping hole it figures. Consider the "soon . . . soon . . . soon" of the novel's conclusion, a conclusion that sends a shiver through me every time that I have read it in recent years. The Creature leaps away to an ice-raft that will "soon" be set on fire. He "shall die" "soon," he tells Walton, and "soon these burning miseries will be extinct," his funeral pile lit, his ascension to it "triumphant." "The light of that conflagration," he continues, "will fade away; my ashes will be swept into the sea by the winds." It is hard not to hear, with a chill, the light of our own conflagration, now currently underway, burning the rain forests of the Amazon, the forests of Australia and Canada, the old-growth forests of California. The angel of history blown into this future, blasting open this moment. In the next sentence this point is underscored in another "hear him, hear him not" formulation: "My spirit will sleep in peace; or if it thinks, it will not surely think thus" (156). My spirit will sleep in peace, the Creature says. *Or.* It may return to haunt us still.

The closing paragraph returns to Walton's voice. I cite it in full: "He sprung from the cabin-window, as he said this, upon the ice-raft which lay close to the vessel. He was soon borne away by the waves, and lost in darkness and distance" (156). He is borne away on the ice raft that he will "soon . . . soon . . . soon" set aflame. I am struck by the way the ice-raft disappears into the darkness, which I read as the darkness of the gaping hole *and* disappears into the distance which I read as our distance, the monster springing forward to land in our future, his ice-raft now barely sustaining his weight, melting fast, the fires we've lit foreshortening that "soon . . . soon . . . soon" once again. For the Creature's predicament is itself a reprise of Frankenstein's own arrival at Walton's ship, and reminds us that we, too, could be "left drifting on a scattered piece of ice, that was continually lessening, and thus preparing for a hideous death" (145). *Or.* We may continue to think but, surely, not think thus.

. . .

PART III

Meanwhile: Two Endings

Meanwhile: First Ending

The Academic Book, Interrupted

I began this book with Walter Benjamin's account of the Tay Bridge train tragedy and the "gaping hole in the line" it produced. I return, in closing, to the Tay River not as a scene of devastation but of possibility. Well before the bridge was built, in the early decades of the nineteenth century, a teenage girl walked its northern shores, whiling away her time. Those shores, she grants, were "blank and dreary" but only "on retrospection." For her at the time, they were "the eyry of freedom," giving rise to "airy flights of . . . imagination" and "true compositions" of the mind. Unconfined by "[her] own identity," she communed with creatures and people who were not only her companions but also her spurs to a sense of self unstitched from the singular (Shelley, "Introduction" 169–70).

That girl on the banks of the Tay River was of course Mary Shelley. In her novel, Frankenstein, too, visits those river banks en route to the remote Orkney Islands where he takes residence to work on his second edition of the Creature, a female companion for the first. But that work, like Walton's journey, will remain "half-finished" (119). And like Walton's journey and Shelley's novel, it invites us to think about what finishing means. Frankenstein walks on the "stony beach of the sea," listens to the waves, and works in a fury of unthinking resolution (113). He wants only to get his work done. Until, "in a pause," he reconsiders (114). He suddenly fears that what he creates "might make the very existence of the species of man a condition precarious and full of terror" (114). Frankenstein senses his obligation to "future ages" and his responsibility to change course.[1] These reflections trigger a fit of "trembling"—the word is repeated four times on one page (115)—as he swiftly destroys what he has created. The Creature, watching from a distance, moves in with new warnings and threats and then leaves in a boat that "was soon lost amidst the waves" (116).

In one of the novel's many slippages of identity, now Frankenstein and not the Creature walks "like a restless spectre" around the island, haunted by the Creature's words that keep returning to him "like a dream" with the heft of "reality" (117). His final act in the Orkneys, however, is not the destruction of his half-finished work, but his erasure of that destruction. Recognizing that it would be unwise to leave any traces of his "odious work," he describes the process of removal as follows:

> The remains of the half-finished creature, whom I had destroyed, lay scattered on the floor, and I almost felt as if I had mangled the living flesh of a human being. I paused to collect myself, and then entered the chamber. With trembling hand I conveyed the instruments out of the room; but I reflected that I ought not to leave the relics of my work to excite the horror and suspicion of the peasants, and I accordingly put them into a basket, with a great quantity of stones, and laying them up, determined to throw them into the sea that very night. (118)

Between "two and three in the morning" he sets off and at a moment when the moon is dimmed by clouds, "cast[s] [his] basket into the sea." "I listened," he writes, "to the gurgling sound as it sunk" (118).

But that act does not rest. It returns. It reverberates outward just as the basket must have left its own encircling waves, receding ever outward, and getting smaller until they were absorbed into the sea. Frankenstein is arrested for murder when he arrives back on shore. This transposition of the drowned remains of a half-finished work, and his friend Henry Clerval, also murdered that night, belies another action that happens and does not happen. Frankenstein does not murder Clerval. And yet he is responsible for Clerval's death. It is hard not to read an allegory, transposed to our current period, of the Anthropocene here. That basket with the remains of half-finished created things keeps arriving on our shores; it traverses time, doubles back, returns, and arrives again. And like the monster itself and angels, ghosts, and other blank blanks, it gestures toward a palimpsestic temporality and perspectives more "acrobatically multidimensional" (Tucker).

As the Global North finally begins to confront its own "storm," perhaps *Frankenstein* can inspire a new storytelling session. Unlike Shelley and her friends who find their motivation in a competition, I imagine instead a storytelling collaboration, a series of braided narratives that, together, seek to chart a "turning around," an approach that is "quite otherwise" and that by definition has not yet been imagined.[2] Those braided narratives, stronger than the sum of their parts, are strengthened by the gaps that define them.[3] This is not a call for finishing the unfinished stories of Shelley's friends but for responding to them, for writing in a register of responsiveness of the sort that Andrew Miller invokes in relation to criticism in general.[4] I don't want to confine my comments to the genre of the novel alone but rather to extend them to all genres and all media (from print to visual to sound to tactile and built cultures). For what I imagine is not only a post-time collaboration between friends and others but also a collaboration of media and approaches that begin to carve spaces in which the "quite otherwise" may flourish.

In recent years there has been an uptick in climate protests and a surge in groups like Extinction Rebellion (in part inspired by Thunberg's words and her deployment of what her own media platform has made possible). They reflect a growing awareness of the gravity of the climate crisis and a growing appetite for social and political change. But change does not happen by protest and policy alone. It also requires the challenging, exciting, and hard work of imagining alternatives. It requires conversations and collaborations, visions tested and rejected and tried again. In short, it requires the vocabulary and the frameworks necessary to make legible and workable a just transition to a global and, most likely stairstepped, fossil fuel emissions elimination plan. To get there we will have to wrestle with and test out possibilities, vie for control of the narrative, and debate the merits and demerits of different positions.[5]

This entire process happens because stories have stakes. Because the work of interpreting the world *is* also one dimension of the work of changing the world. And the work of interpreting the world is also indelibly bound up with changing the time. If Shelley and her friends sought specifically to create the supernatural effects they admired in the ghost stories of others, they also identify, in that goal, the supernatural effects that are the result of all storytelling: the work of changing or adding to one's perception to the point where one sees what one did not see before.[6] This process could be understood as a certain sort of ghost raising, or perhaps better, spirit raising—although in this case the medium would not be the leader of a séance but the form of the debates, the stories that are told, where, how, and by whom. Supernatural effects, in this sense, are what we often want when we seek to communicate with others. But they are rare: much more typical are the consolidating effects of conventional norms. Which is not to say that this work is not important too. It upholds the values that were once introduced through supernatural effects—values like equality and justice—and maintains them, typically with many small and large adjustments, over time. After a once-contested issue gains public assent, in other words, it becomes absorbed into the way a community thinks. There is, to be sure, something to be said for not letting ideas settle into conventional norms, for asking questions, considering the work ideas do, and whether one wants to continue to endorse them or, instead, agitate for change again.

The climate change idea is still in that vital and volatile process of being worked out. It may have been discussed in the nineteenth century, but the global threat it poses did not begin to be clearly articulated until the 1980s and 1990s. This process of shaping the idea—a process that is also bound up with shaping a response—is ongoing.[7] Many of this idea's key components have been integrated into daily life, have become so familiar and second nature that they do not need to be named (in my city, these ideas would include recycling, composting, and gardening, to note a few). And many have been named in a way that makes us see something that we did not see before (in my city, the introduction of the UV index into weather reports is one example). But the very fact that I note *in my city* underscores that these shifts are not yet widely established. And the piecemeal shifts that have taken place tend to be the small, everyday things I note above, which, to be sure, do require some infrastructure but are mainly about preparing the way—building the mental muscles, instilling habits, inciting interest, creating networks and movements—for the real and lasting change that counts. Preparing the way, that is, for the sort of change that is the difference between human and other species's survival and human and other species's extinction. In this context, the debate is still very much alive and unresolved. My call here to stay with the gaping hole—to hold the line and open the line—is a call to envision alternative temporalities better equipped to address the crisis in a spirit of conversation and collaboration, experimentation and failure, and modes of care that forge connections across and through time rather than heeding only the timeline with its sights set on "the end."

It may seem odd that I turn to a novel from the early nineteenth century to frame these comments. But because Shelley's novel is so keenly attuned to framing and because it throws into such nice relief what the humanities can contribute to the climate crisis—not only articulating the terms through which the climate crisis may be addressed but also shining a spotlight on that process and showing us how it works—it offers fertile ground for further responses to climate disruption. Turning to the nineteenth century, as I have done with Shelley and many others in this book, however, also does something else: it activates the discordant, overlapping temporalities that I have been arguing are necessary to navigate if we are to have an adequate grasp of the climate response required of us now. Not only does *Frankenstein* itself mobilize these discordant temporalities; by putting this novel into dialogue with our current climate moment, it also reminds us that we need not remain invested in those journeys of progress by which Walton and Frankenstein were enamored. It reminds us that we can instead better act from a temporal modality alert to what can be learned from the disjunctions and overlaps. When one maintains the gaping hole, as Shelley does, what emerges is the time-bending co-written frame of letter writing. Walton's letters to his sister and to us are porous, collaborative, and community-creating post-time missives that arrive, already opened, in our hands.

I began this book thinking about how the typical ways in which the Global North formulates problems in an effort to mobilize action (as representation or warning, among others) have failed to make a dent in carbon emissions.[8] As a nineteenth-century scholar, I went to the works I knew, hopeful that they might offer routes out of that impasse. Turning to these works, by some accounts, was also turning to the origins of the Anthropocene, and that felt fitting. But as I continued to research and write, and as I continued to confront daily my own perplexity with the failure of convincing description to generate climate action (the sort of description that steals our souls and hearts, makes us weep, makes us look up at the world and see it with fresh eyes—the sort of description we so often get, in short, in literature), I encountered something that I didn't expect. I encountered the many moments in nineteenth-century literature and culture when people did not know what to do, when they were themselves confronted by events that exceeded their capacity to frame them, when they were frightened or wary or full of disbelief.[9]

And what I found there was a skip in the record. The anachronistic metaphor is intended: it is the embodiment of post-time as material, as palimpsestic, as polyphonic and polytemporal. The skip is the place where the vinyl is marred, where the song or music or speech is compromised, less perfect than it should be. And I realized something that we all know but do not always emphasize: that our stories straighten things, they include some details and exclude others. I've accordingly read the works that I address in this book attentive to the stutter and the stumbles, the places where meanings falter, where the bridge threatens to give way. I've tried to remain alert to those times when we, as critics, introduce reinforcements—iron girders, tighter floor boards, heavier concrete anchors—to subdue the sway and the tremor that the works otherwise produce as they list in the wind of our interpretations. I've tried to resist nailing them more firmly in place.

The question that underpins Shelley's book also underpins mine: How do we act—on anything—when we are looking straight at a gaping hole, be it a monster or climate change or any other thing that defies our capacity for representation? And yet act we must. As I write in the solitary space of my study, I am struck by the many collaborative projects I have recorded in this book: not just Shelley's storytelling sessions with her friends but also the collaborations that comprised Henry Mayhew's *London Labour* and Richard Mosse's *Incoming*, not to mention the ongoing collaborative project of climate action both on the streets and in global cultural projects of astonishing diversity and range. Not to mention, too, my debt to Virginia Woolf and Walter Benjamin and their sustained attention to co-writing as an act of defiance and time-crossing alliance. Indeed, we are always collaborating and co-writing, entering into the ongoing scholarly and cultural conversations that rearrange time and the way we occupy it even if we don't name, or even notice, such rearrangement as such.

Which returns me to Ann Cognito. In opposition to singular identity and societal expectation, she pitched a tent on a traffic island in Ottawa. Soon others followed and began conversations in which strangers huddled close together in a shared project of social justice. One of my friends, a doctor who had not been a climate activist until very recently, said to me one day, "I'm a climate kook." He meant that he was ready to do whatever it took to address climate change. He meant that most of his friends, and even his family, were going to think he was, as he put it, a "kook" (Creaghan, 13). He meant that he entered Cognito's tent, stayed overnight, joined that small but growing band in climate protest. If *kook* means "mad or eccentric person," we also know that madness and eccentricity are defined according to the terms a society dictates. Kook itself derives from *cuckoo*, that bird whose iterative cry is captured in the name itself, and who lays its eggs in the nests of other birds. Long interpreted in a derogatory way in relation to private property—Don't put your eggs in my nest—I wonder about seeing the cuckoo's practice as a communal invitation to harmonious co-living.[10]

But what does this have to do with temporality? The cuckoo clock was invented in the mid-eighteenth century and continues to adorn homes today; it spans the period of this book. In these clocks, a wooden cuckoo bird calls the hour on the hour. But instead of containing the cuckoo in the clock, can we imagine cuckoos and climate kooks otherwise? Can we begin, with them, to occupy time differently and find new ways of telling the time? This will mean attending to Indigenous traditions and knowledges that have been for too long neglected in approaches to climate change in the Global North. It will also mean, as I have done here, reinflecting and co-writing the traditions and knowledges of the long nineteenth century and our own period in the spirit of climate action collaboration, finding and experimenting with new forms, and remaining open to new possibilities. For what could be more mad and eccentric than to stay the current course and continue to confront the crisis with tools that demonstrably do not work? Let's unfold the figurative tent—be it writing, activism, science, or some interdisciplinary combination of all three—set up camp, find new ways to occupy time, and, through doing so, change the time that has brought us here.

. . .

Meanwhile: Second Ending

Unfinished

. . .

Let's attend the ellipses. Let's unfold the figurative tent—be it writing, activism, science, or some interdisciplinary combination of all three—set up camp, find new ways to occupy time, and, through doing so, change the time that has brought us here. There is always more than one ending, and even when they sound the same, as does the cuckoo's cry, their repetition can make all the difference, opening folds in time that we have not yet begun to imagine.

. . .

Acknowledgments

I began this book in 2015 when the climate crisis was only occasionally in the news where I live, and I finished it amid heat domes, atmospheric rivers, and floods. Yet no work has such neat boundaries. This book's origins pre-date the moment I began writing it and, like all books, its ends recede to unpredictable, unbounded reading horizons. So, too, its author. While I am the nominal author, my debts are far and wide, blurring my own boundaries, and bringing into focus instead the many people who have informed, sustained, and inspired my thinking and writing over the years. It is a pleasure to thank them here. I want to acknowledge, too, the Algonquin peoples on whose unceded land I wrote this book.

This project has been buoyed and supported by many groups. As anyone who engages with the climate crisis knows, communities are a prized gift in helping to parse the day-to-day, to appreciate what collective work makes possible, to manage dismay and disbelief, and, equally, to share joy and laughter. The dedication, generosity, and wisdom of the people with whom I've been lucky enough to work has been a tonic against the inevitable difficulties that are part of any writing project and that are only deepened when that project confronts the climate crisis. The Carleton Climate Commons informs this book at every step. I am deeply indebted to Vladimir Díaz Cuéllar, Danielle DiNovelli-Lang, Lenore Fahrig, Veronika

Kratz, Alicia Myc, Franny Nudelman, Justin Paulson, Chris Russill, Kim Siguoin, Peter Thompson, and Brenda Vellino, and to the many colleagues, students, and administrators at Carleton who have come together since the Climate Commons' inception in 2014. Chris Russill, in particular, has been a catalyst for my thinking in more ways than I can document here. Special thanks, too, to Noel Salmond, whose casual remark in one of our early meetings planted the seeds for this book.

Carleton University, overall, has offered me a remarkably congenial academic home over the past thirty years. My colleagues in the English Department and the Institute for the Comparative Study of Literature, Art, and Culture have been unfailingly supportive and model what the best forms of community can look like. Janice Schroeder and Pascal Gin, chair of English and director of ICSLAC respectively, and Pauline Rankin, dean of the Faculty of Arts and Social Sciences, are all administrators extraordinaire who make their demanding roles look effortless and who have never, in my experience, lost their humor, compassion, and superpower capacities to surmount skyscraper-like obstacles in a single leap. Through the Climate Commons I have also come to know the chairs, directors, and faculty members of many other departments with which I am not officially affiliated but to whom I am grateful for their quick and unqualified support for anything our institution could do to contribute to climate action. I also want to thank my fellow co-instructors in our new interdisciplinary graduate course on climate change: Patrick Callery, Danielle DiNovelli-Lang, Alex Mallett, James Meadowcroft, Ron Miller, Kristen Schnell, and Jesse Vermaire, all of whom gave me an inspiring vision of what interdisciplinary climate change pedagogy can look like.

Like the news cycle, my students' sense of the climate crisis has transformed over the past five years. I am grateful, and inspired, every day for what they teach me about living in conditions of climate precarity. They didn't always know what they were signing up for, but they stepped up nevertheless and, again and again, produced dazzling insights on the climate crisis. Many thanks, in particular, to Mariam Abdallah, Emma Andrigo, David Baker, Clayton Budge, Emma Cantlon, Skylar Chambers, Taissa Cronin, Rayan Eid, Brayden Last, Phillipa MacDonald, Mary Petruschke, Sam Taylor, and Jack Wilson. I am also grateful to Danielle Taylor for transforming my messy Works Cited into (relatively speaking) a thing of beauty. A quartet of graduate student research assistants helped me with this book in countless ways; I want to thank, in particular, Sara Adams, Jenna Herdman, Veronika Kratz, and Gemma Marr. They exemplify brilliance, commitment, and kindness in these times.

Many people read chapters, made suggestions, and/or supported this book in myriad other ways. I am grateful to Sara Adams, Kristin George Bagdanov, Carolyn Betensky, Tina Young Choi, Jason Creaghan, Rachel Fetherston, Susan Stanford Friedman, Anna Henchman, Stephanie Lemenager, Ben Lerner, Anthony Lioi, Sarah Luria, Kyveli Mavrokordopoulu, Richard Mosse, David Pare, Karen Pinkus, Matthew Purvis, Samia Rahimtoola, Anne Raine, Jonathan Sachs, Jerry Singerman, Stephen Siperstein, Kelly Sullivan, Luang Por Viradhammo, Priscilla Wald, Chris Walker, and Samuel Weber. I am especially grateful to Deborah Morse and the other members of the "Support Group," Mary Jean Corbett, Teresa Mangum, Ellen Rosenman, and Talia Schaffer, all of whom so beautifully model scholarship, friendship, and living.

This book has also benefited enormously from the feedback I received at several conferences: the Interdisciplinary Nineteenth Century Conference, the National Victorian Studies Association Conference, the Northeast Victorian Studies Conference, the CUNY Graduate Centre "Victorian Ecotime" Conference, the Victorian Studies Association of Ontario Conference, and the Vcologies Conference. The Vcologies group has been a source of inspiration and information and a wonderful forum for lively debate since its beginnings. Special thanks, in particular, to Siobhan Carroll, Dennis Denisoff, Kate Flint, Kathleen Frederickson, Eric Gidal, Devin Griffiths, Nathan Hensley, Deanna Kreisel, Tobias Menely, Richard Menke, Liz Miller, John Miller, Benjamin Morgan, Jesse Oak Taylor, Michael Tondre, Daniel Benjamin Williams, and Lynn Voskuil.

I could not have asked for more incisive, wise, and generous comments on this book manuscript than those offered by its two anonymous readers. Their suggestions dramatically improved this book. I am deeply grateful.

Several friends provided much-needed days and evenings of walking, talking, cycling, skiing, eating, drinking, and larking. To all of the members of the Diaspora group who so easeasily transitioned from wacky New Year's Eve line-dancing to Zoom squares, I am grateful for your camaraderie and support: Rachelle Abrahami, Michael Berkowitz, George Bourozikas,

Rachel Kirtner, Hillary Kunins, Gordon Lafer, Amy Shire, Peter Sidi, Photini Sinnis, and Marla Stone. Warm thanks, too, to Sandra Flear, Louise Gwyn, Peter Halewood, Peter O'Brien, Mary Stinson, and Donna Young. I also could not have written this book without the support and kindness of my Ottawa friends, many of whom are also colleagues: Parker Duchemin, Mitchell Frank, David Holton, Jodie Medd, Franny Nudelman, Jan Schroeder, Carol Shepherd, Laura Taler, Richard Tarasofsky, and Brenda Vellino. Franny and David cooked many meals that sustained this writing, eased pandemic isolations around an outdoor fire pit in the winter, and offered distractions when they were most needed. Brenda, Jodie, and Jan, in addition to providing food and friendship, read and responded to several sections of this book manuscript with wonderful wisdom, a keen eye for my infelicities of phrasing, and a canny sense of the possible.

This book would not have been written without the Insight Research Grant and Institutional Grants I received from the Social Sciences and Humanities Research Council of Canada. Nor would it have been written without generous support from Carleton University. I am grateful to both institutions.

Heartfelt thanks also to Stanford University Press. Faith Wilson Stein's wise input over the three-plus years of its writing were hugely appreciated. Caroline McKusick steered this book through its final stages with skill and aplomb. And Erica Wetter provided a transitional bridge that ensured we stayed on course. I am deeply grateful they were willing to take a chance on the strangeness of this book.

At every step along the way my immediate and extended family has provided support, laughter, and feedback on my work (whether I wanted it or not). I am so grateful for my siblings, Steve (for his craftsmanship, photography, and vegan baking), Ted (for his computer programming mantra, "Fail quickly," and his psychotherapy mantra, "Many things . . ."), Ann (for being there when I need her), and my siblings-in-law Miriam and Joel (for family get-togethers and funny jokes). My mother, Rosemary Leckie, is an artist and creator extraordinaire, and she taught me to see the world slant. My late father, Robin Leckie, insisted always on acting from one's convictions and his sense of justice continues to inspire my writing and thinking every day. My mother-in-law, Ruth Westheimer, is a model of "do it now" thinking, and if the world could apply her thinking to the climate crisis, we'd all be better off. Judith Leckie, my aunt, read the book manuscript not once but twice. Her lively and spirited comments compelled me to revisit and rewrite sections because she was (almost) always right. The book is much better after her input, although, as she would be quick to remind one, all the errors that remain are my own. When the limitations of Word's formatting defaults made it difficult to write the "Layering" chapter, my nephew, Noam Flear, said, "Let me write a computer program to fix that." And he did. I am immensely grateful. Huge thanks, too, to Robbie Rose Landry, Maeve Leckie, Alannah and Alicia Frick, Ari and Leora Einleger, and Isaac Kinley, all of whom brought pleasure, joy, and downtime to my life while also reminding me of how the climate crisis *feels* when one is young.

Last and most, my immediate family: Michal Leckie, now a graduate student in Global Health, routinely sends me articles, communicates ideas, and responds to my own work in ways that flex and expand my thinking. Wisdom fount, accidental comedian, and strict editor, I am deeply grateful for her input. Ben Westheimer is a talented media producer, musician, and burrito maker, and a perceptive student of the state of the world; his offbeat insights inform my thinking always. Joel Westheimer adds a zest and zip to each day, and every page here owes a debt to him. He is my co-, my love, my companion in all things, and I am incredibly grateful for our life together.

As I was writing this book, I often looked at the view from my window. Often, too, I closed my computer and went outside. I would be remiss in not acknowledging a debt, too immense to express, to the great beauty of the world that is always there, even on the bleakest days.

Notes

PREFACE

1. In making this distinction between climate change and the climate change idea I am drawing, in part, on Edward Said's distinction between the Orient and Orientalism (2–3). See also my discussion of the architectural idea (*Open Houses*, 37–42) and the sanitary idea ("Introduction," *Sanitary*).

2. Tobias Menely makes the point concisely: the Anthropocene is "a name for a problem of time" ("Ecologies," 85). See his "Ecologies of Time" for an excellent overview of the ways in which critics have addressed questions of time in relation to climate change. See also Elizabeth Callaway, Dipesh Chakrabarty, Anne-Lise Françoise, and Bruno Latour. Ian Baucom offers the most ambitious response to these issues that I know of to date.

3. See also Callaway, Wendy Hui Kyong Chun, Timothy Clark, Amitav Ghosh, Timothy Morton, and Rob Nixon among many others. As Clark puts it, climate change "resist[s] representation at the kinds of scale on which most thinking, culture, arts and politics operate" (x).

4. The research in time studies is vast, and I cannot come close to covering the rich material that has contributed to my thinking in this area. For a philosophical overview, I found Hoy especially helpful. His book does not replace reading the philosophers themselves, but it does provide a lucid overview of the different philosophical positions on time spanning the Greeks to our current moment. The philosophers most helpful to this project were Giorgio Agamben, Jacques Derrida, Elizabeth Grosz, Martin Heidegger, Bruno Latour, and, especially, Walter Benjamin. For considerations of time in relation to the long nineteenth century, Nathan

Hensley, Tobias Menely, Benjamin Morgan, Jesse Oak Taylor, and Sue Zemka (all of whom also address climate change) were indispensable. Elizabeth Miller's *Extraction Ecologies* was published after this book was completed but it, too, addresses nineteenth-century studies, questions of time, and climate change with great insight. The revelatory work of Susan Stanford Friedman helped to orient me in relation to modernist studies. For treatments of contemporary temporalities, two edited collections were especially useful, Amelia Groom's *Time: Documents of Contemporary Art* and Joel Burges and Amy J. Elias's *Time: A Vocabulary of the Present*. Two fields that offer rich treatments of time are queer studies and Indigenous studies. Here Elizabeth Freeman's work is pivotal in relation to queer studies and Nick Estes's, Robin Kimmerer's, and Kyle Whyte's work offered me roads into Indigenous studies. I want to stress that none of the ideas I explore here are new; they have all been covered in inventive and compelling ways by others, but they have not always been rallied together for climate change action. In part, then, this book seeks to revive older ideas and put them into dialogue with each other and with the climate crisis. While I build, in particular, on existing studies that consider Benjamin's contribution to time studies and to climate change in a largely Western tradition, I have come to believe that Indigenous studies now offers the most supple, generous, and capacious orientation to time, and I hope future studies will further explore what I have attempted to begin in this book. Each of the references here also includes extensive bibliographies that are worth consulting. Finally, I argue throughout that narratives of progress become entrenched with the Industrial Revolution and, for the most part, have only deepened since then (see Donna Haraway, Latour, Anna Tsing, and Imre Szeman in addition to many of the critics noted here).

5. See Marco Caracciolo, Choi and Leckie, Jonathan Sachs, and Sue Zemka for elaborations of slow time. Interestingly, the timeline itself only began to take hold in the public imagination in the later eighteenth century (Daniel Rosenberg, 60).

6. There was extensive newspaper coverage of the accident in the period. These details come from local newspapers as well as the *Illustrated London News* and Mr. Rothery's lengthy courtroom report.

7. A comment on my title and subtitle: the phrase "Climate Change, Interrupted" will be familiar to some readers in the context of Vermeer's "Girl Interrupted at Her Music" (1660–61); and to others in the context of Suzanne Kaysen's account of mental illness, *Girl, Interrupted*; and to yet others still in the many works that adopt this locution. The first chapter of Eugene Richardson's *Epidemic Illusions*, for example, is called "Colonizer, Interrupted." My own reasons for choosing this formulation are many. All titles work in unexpected ways, encouraging some comparisons and discouraging others, and I see this title as part of a wider conversation that can be pursued in several directions and that carries a certain history within it. The subtitle was added *after* the book was written, as subtitles so often are. I could comment on the time of writing and how the introduction of a subtitle changes what has already been written, but I will

limit myself to noting only that had I been writing with this particular subtitle in mind, I would have engaged more directly with the debates on representation (and description, and realism, and narrative) in my field. As will be clear in the pages to follow, I use the term *representation* capaciously and variously, and I hope its meaning will always be clear from the context provided. The "re- . . . re-" in my subtitle also nicely captures the complexity of time—what is anew *and* returned, what is behind *and* after.

8. In general, I use "we" and "our" to refer to communities of academics, although sometimes, as here, I extend the reference to embrace a shared narrative in the Global North. That said, Sylvia Wynter's work has alerted me to the pitfalls and assumptions of what she calls the "referent-we" (Wynter in McKittrick, 24; see also McKittrick, 7). The referent-we defaults to unnamed but assumed privilege while implying that it includes everyone. As a result, the referent-we makes silent exclusions that Wynter seeks to make visible and, in the process, to transform.

9. See Ailton Krenak's *Ideas to Postpone the End of the World*, Bill McKibben's *End of Nature*, Timothy Morton's *Hyperobjects: Philosophy and Ecology after the End of the World*, Roy Scranton's *Learning to Die in the Anthropocene: Reflections on the End of a Civilization*, and Slavoj Žižek's *Living in the End Times*, among others.

10. See especially Gould as well as the critics in note 4. It is worth noting that James Secord does not agree with Gould's interpretation of Charles Lyell.

11. See Janice Lee's interview with Robin Kimmerer for a discussion of the power of not naming.

12. See Kari Marie Norgaard and Alan MacDuffie ("Charles Darwin") on climate grief and climate denial. See Andri Magnason for an account of the ways in which language often fails to cohere into something that makes sense in the context of climate change; he suggests that many people often hear, instead, "white noise" because climate change concepts are inadequate to the enormity of the crisis.

13. See Jonathan Crary's *24/7* for an excellent discussion of how late capitalism redefines the meaning of free time.

14. Christina Sharpe, drawing on Dionne Brand, considers "sitting with" as a method for Black studies (13). I will discuss "sitting with" in the context of interruption in the first Beginning, "Interruption."

15. This sense of interlocking and connected crises that extend beyond the climate crisis and the pandemic to include anti-BIPOC racism, the decline of democracy, and economic inequality has been discussed perceptively by a number of critics. See, for example, Dionne Brand, Arundhati Roy, and Kim Tallbear. See also my six-part podcast with Joel Westheimer. I am indebted to Chris Russill for the phrase "same emergency."

16. At the same time, the fracture lines in communities around the world also came into vivid relief. As added stress was placed on unequal and unrealized infrastructures, it became impossible to continue to uphold these structures without public outcry, protest, and calls for reform. The interruption the pandemic

produced may perhaps be understood as an example of what happens when a gaping hole in the line comes into view and is not sutured over.

17. See the January 2021 issue of *The Lancet* entitled "Climate and Covid-19: Converging Crises."

INTERRUPTION

1. These origins are variously defined in terms of the introduction of agriculture (10,000 years ago), the impact of colonial exploration (1610), the invention of the steam engine (1784), or the great acceleration (1952–54).

2. See the Appendix to Lemenager's *Living Oil* (197–200).

3. Some might respond that e-books would offset many of these objections, but they have also been debated in this context.

4. My understanding of "thinking" here embraces both thinking in the everyday sense (thinking about things, mulling over questions, pursuing ideas) and thinking in the Benjaminian sense. In *The Arcades Project*, Walter Benjamin famously describes thinking as follows: "Thinking involves both thoughts in motion and thoughts at rest. When thinking reaches a standstill in a constellation saturated with tensions, the dialectical image appears. This image is the caesura [i.e., interruption] in the movement of thought. Its locus is of course not arbitrary. In short it is to be found wherever the tension between dialectical oppositions is greatest. The dialectical image is, accordingly, the very object constructed in the materialist presentation of history. It is identical with the historical object; it justifies its being blasted out of the continuum of the historical process" (475).

5. A lot of the early work in this area came from science and technology studies as well as the social sciences. Latour's lament, in 1990, that bringing the different disciplines into dialogue and recognizing their interconnection was "unthinkable" no longer holds—there has been a lot of work and a lot of thinking on precisely this point—*but* there is also certainly more that could be said. "Our intellectual life is out of kilter. Epistemology, the social sciences, the sciences of texts—all have their privileged vantage point, provided that they remain separate. If the creatures we are pursuing cross all three spaces, we are no longer understood. . . . In the eyes of the critics the ozone hole above our heads, the moral law in our hearts, the autonomous text, may each be of interest, but only separately. That a delicate shuttle should have woven together the heavens, industry, texts, souls and moral law—this remains uncanny, unthinkable, unseemly" (Latour, *We Have Never Been Modern*, 5).

6. They also remind us that thinking itself is not reserved for humans and help to shift us out of an anthropocentric frame.

7. See also Estes's "Water Protectors" and Tallbear for discussions of practice as theory.

8. There are many books that wrestle with what the humanities can contribute to climate change action, with the divide between science and the humanities,

and, to a degree, with explorations of how the way in which a problem is defined shapes the solutions that can be imagined. The humanities are keenly attuned to how language works; humanities critics point out, for example, that in the phrase *climate science* "climate" often rubs up against "science," undermining its authority. And there have also been robust discussions of the pros and cons of the terms "global warming," "climate change," "climate crisis," "climate emergency," and "ecocide," among others. See especially Hensley and Steer.

9. See also Tsing's latest collaborative project, an interdisciplinary and interactive website, *Feral Atlas* (Tsing et al., 2021).

10. There is, of course, a long and sometimes vexed history of formal experimentation in relation to the academic book. Jacques Derrida's work, especially in the latter half of his career, perhaps offers the best example here. In a more popular context, Jonathan Safran Foer's *We Are the Weather* (2019) experiments with form, with varying degrees of success, through lists, a staged-debate format (echoing a Socratic dialogue), and a personal voice. See also Jeffrey Jerome Cohen's and Steve Mentz's essays in Menely's and Taylor's *Anthropocene Reading* and Susan Stanford Friedman's "Scaling Planetarity," which she calls a "collage of meditations" (119). See also Eric Hayot's lively essay on academic writing.

11. See Derrida's discussion of the translation of *je-weilig* in Heidegger as at once "lingering awhile" and "a passage . . . whose transition comes, if one can say that, from the future" (*Specters of Marx*, 28).

12. Pierre Bourdieu similarly notes how visual art at the end of the nineteenth century produced new ways of thinking that disturbed people's minds (148–49).

13. I'm indebted to Samuel Weber for drawing my attention to Brecht's role in Benjamin's thinking about interruption. (I have used the John Heckman translation here.)

14. Alan Weisman (2007) introduces this term that is now oft-used.

15. See Chakrabarty's *The Climate of History in a Planetary Age* for an expansion and extension of this essay and a careful response its critics.

16. Two excellent recent books—Jason Groves's *The Geological Unconscious* and Tobias Menely's *Climate and the Making of Worlds*—consider earlier literary periods through the double lens of Benjamin and climate change, but they don't discuss the angel of history.

17. Ferris stresses "how important Benjamin's style of writing is to his thought" (3); his work "squarely relocates thought in the means of expression rather than seeing it as something that expression represents" (8).

18. See also Andrew Benjamin, Samuel Weber, Michael Jennings, and Buck-Morss in particular.

19. See especially Weber's discussion of Derrida and Benjamin in this context (122–28).

20. As Ghosh writes, "at exactly the time when it has become clear that global warming is in every sense a collective predicament, humanity finds itself in the thrall of a dominant culture in which the idea of the collective has been exiled

from politics, economics, and literature alike" (80). Scranton, Klein, Latour, and many others make the same point.

21. See Tobias Menely and Jesse Oak Taylor for a good summary of the debated origins for the Anthropocene (1–25).

22. I am indebted to Devin Griffiths for this suggestion (personal correspondence).

23. Here one could argue, for example, that sustainability discourse is an outmoded form of thought insofar as it cleaves to a capitalist framework in a moment when that framework only functions with a commitment to unrestricted growth that cannot be reconciled with a reduction in carbon emissions. This example is obviously open to debate, and others would offer different examples.

24. Christina Sharpe uses this language of "sitting with" to think about methodology in a way that also chimes with my treatment of post-time in this book: "I've been trying to articulate a method of encountering the past that is not past. A method along the lines of a sitting with, a gathering, and a tracking of phenomena that disproportionately and devastatingly affect Black peoples any and everywhere we are" (13). One of those phenomena that disproportionately and devastatingly affect Black peoples is, of course, climate change.

25. Just as a rush of stories "do not nest neatly" (Tsing, *Mushroom*, 37), so, too, the timescales of the climate crisis do not nest neatly.

26. A. Benjamin writes: "The caesura allows . . . the relationship between the particular and the Absolute to be thought" (104–5).

27. Weber later writes: "But if it is a way, if it makes a way, where is it headed? Not simply back to the original or to the origin, but rather *away* from it. In moving away from the original, translation unfolds *the ways* of meaning by moving words *away* from the meanings habitually attached to them, and which are generally construed as points of arrival rather than of departure. Meaning is generally conceived as a self-contained, self-standing universally valid entity, one that precedes the words that express it. Translation's way to go, by contrast, leads in the direction of other words and meanings, exposing a complex and multidimensional network of signification in which word-occurrences are inevitably inscribed" (92).

28. There are two titles for this work, speaking again to the vagaries of translation.

29. Callaway offers a good reading of messianic time and climate change through the prism of Benjamin and Agamben: "Instead of defeatist attitudes that may result from climate change presented as apocalypse, or as too slow and banal to represent, messianic time is a call to action, a call to have a different relationship to time, even distant time. We are urged to recognize that the past has great bearing on the present, we are called upon to take up the monumental task of imagining potential worlds, and we are challenged to recognize that we create these worlds in our present" (31).

30. The *Arcades* passage cited earlier is, of course, a rephrasing of this citation.

31. I will return to deep time in "On Layering."

32. On the image, in particular, Stephanie Lemenager writes that environmental discourse's "attraction to middle-class rhetorics of rights, consumption, and sacrifice forecloses structural critique, and its overinvestment in spectatorship [oil-covered birds] troubles its relationship to action" (*Living Oil*, 24). Sontag, Ranciere, and Nelson, she continues, "theorize the deficiencies of 'the image' as a route to action—with images figuring as 'our normal condition,' in Ranciere's words, rather than a special prompt to social change" (35).

33. It also contributes to the superimposition or binding together of Protestantism, Catholicism, and Judaism in Benjamin's eighteen theses. The Angelus Novus is drawn on an image of Luther, thus placing Benjamin's theses into dialogue with Luther's; in Catholicism the Angelus bell is rung eighteen times as a call to stop what one is doing and remember the incarnation of Christ; and in Judaism the number eighteen represents being alive, or life, through the Hebrew *chai* (life) whose letters add up to eighteen. The Amidah, moreover, was initially composed of eighteen blessings.

34. The temporal models at play in different Indigenous times are also relevant here. Indeed, Indigenous thinking resonates with, but of course is also very different from, how I develop Benjamin's reflections on time for the current moment.

35. There are many different ways this story could be told. Time studies is rife with examples of innovative treatments of time, from surrealist experiments through modernism and postmodernism to our current moment. Visual culture, in particular, offers rich explorations of alternative temporal modes (see, for example, Amelia Groom's *Time: Documents of Contemporary Art*). But if these works push back on chronological or linear temporal modes, there are also several models for living time that have never subscribed to the temporal shift in the Global North following industrial modernity. Many Indigenous cultures, for example, have temporally polyphonic ways of living time from which the Global North has much to learn. See, for example, Nick Estes, Kyle White, and Robin Kimmerer. I focus on Benjamin here because he has so often been a touchstone for climate change scholarship, because his thinking on the dialectical image spans the period I address, and because his work offers formal innovations that are suggestive for academic scholarship in climate change studies.

36. Ferris also writes: "Although this rescue [of the past from conformity] is continual, it is not continual in the sense of an unbroken line but rather in the sense of what has to be repeated over and over again, what in fact has to be begun over and over again" (15).

POST-TIME

1. Of course Eliot is not alone in doing so. Jonathan Sachs notes Wordsworth's sense that "rapid communication has dulled the audience for poetry circa 1800" ("Slow Time," 316) and is linked to "the increasing accumulation of men in cities, where the uniformity of their occupations produces a craving for extraordinary

incident, which the rapid communication of intelligence hourly gratifies" (cited in Sachs, "Slow Time," 317). And later in the century, to counter the accelerated tempo of late nineteenth-century media, James encouraged his readers to read his work aloud—in sotto voce, as he put it—to ensure that they would be forced to slow down (1108). Most recently, Jonathan Crary has taken up similar issues in *24/7*.

2. Periodicity derives from the Greek *periodikos* and indicates the "coming round at intervals," which itself derives from *periodos* (orbit, recurrence, course) and can be further broken down as *peri* (around) and *hodos* (way, course). See Cohen for an innovative discussion of the period in relation to the Anthropocene.

3. It is interesting that the early 1990s is the date that many scholars name as the beginning of serious thinking on climate change. Chakrabarty writes, for example, that "self-conscious discussions of global warming in the public realm began in the late 1980s and early 1990s, the same period in which social scientists and humanists began to discuss globalization" ("Climate of History," 198–99).

4. See Heidi Scott's *Chaos and Cosmos* and Eric Gidal's *Ossianic Unconformities* for two studies that illustrate what we can learn about nonlinearity from eighteenth- and nineteenth-century culture. Naming is an important factor here too. As Jason Moore writes: "If naming can be a first step to seeing, it is also more than a discursive act. In the circumstances of civilizational crisis, as the old structures of knowledge come unraveled without yet being interred, the imperative and the power of fresh conceptual language can become a 'material force,' as Marx might say" (4).

5. See note 1 in the Preface.

6. Eliot's famous consideration of the Dutch painters is here, for example. See George Levine's *The Realist Imagination* for a discussion of realism as a genre with conventions of which writers were acutely aware.

7. Yeazell notes, tellingly, that this passage begins with a lady who interrupts (she calls her the "resisting reader") (91).

8. Pastime itself, of course, defines leisure; here I am trying to invoke both the past, time passing, and leisure together in the concept of past-time.

9. Here I build on what Jonathan Sachs and Sue Zemka refer to as "slow time" and what Tina Young Choi and I refer to as "slow causality." Eliot's narrator in *Adam Bede* illustrates one moment in the history of time when its tempo changed and became marked by, as Ian Duncan puts it, "an accelerated temporal economy" (472). See Jameson's "End of Temporality" for a discussion of the "sense of an alternate temporality" to which living "in two distinctive worlds simultaneously" gives rise. His example is the same as Eliot's: the brief period of straddling agricultural temporal rhythms and lives and modern, industrial ones (699).

10. Notably, this new word for time's palimpsestic quality, its mediation, and its now accelerated tempo was coined several decades before the period that John Guillory identifies with mediation's return to visibility via the telegraph, telegram, and telephone among other new technologies ("Genesis").

11. Elizabeth Freeman captures this point nicely as follows: "Chrononormativity is a mode of implantation, a technique by which institutional forces come to seem like somatic facts. Schedules, calendars, time zones, and even wristwatches inculcate what the sociologist Evitar Zerubavel calls 'hidden rhythms,' forms of temporal experience that seem natural to those they privilege" (*Time Binds*, 3).

12. Benjamin Morgan also opens up the possibilities of "after" in "After the Arctic Sublime." He writes: "It might seem natural to think of our temporal relationship to nineteenth-century Arctic tropes in linear terms: we are 'after' the Arctic sublime in the sense that the twenty-first century is after the nineteenth, in the sense that the global climate warms year after year. But the following pages explore a different kind of afterness, one that can be recovered within certain texts as having been there all along" (4).

13. This idea of after-time invites a comparison with afterlives as discussed by Christina Sharpe in *In the Wake*. In this context, I am indebted to Nathan Hensley's INCS panel on Sharpe's impact on Victorianist thinking.

14. We now live in a world of many *posts* where the "post" modifies the word it precedes. These posts signal updates of categories that require expansion. The post maintains that earlier category but revises it. In most cases, however, the update is not a wholesale dismissal of the earlier category even when the "post-" reverses or eviscerates that earlier category's founding assumptions. I value and often use many of these "post" formulations; at the same time, I am wary of the work they sometimes do.

15. See Mullen's *Novel Institutions* for an exceptionally penetrating treatment of anachronisms in nineteenth-century narrative. See also Freeman's "Synchronic / Anachronic."

16. Castronovo, for example, notes that Benjamin describes his concept of "alternative historical sensibility" as "a 'constellation' that has the potential to fuse the past and present into a single formation. Any method that could be adduced from this philosophy of history would likely be grounded in anachronism, which is hardly stable ground at all, since it jumbles the logic of chronology. 'Anachronism is the mistake that consists of putting a fact too early,' writes Ranciere. It is also the mistake that unmoors facts from the narrow purview of dates and allows them to resonate beyond their immediate temporal locations" (1145). See also Mullen and Derrida (*Specters of Marx*, 25–26).

17. For early reviews of *London Labour*, see Mayhew, *London Labour and the London Poor: Selections*, 311–18. Some of these reviews, from the streetfolk themselves, notably take issue with his methodology.

18. The latter two volumes of *London Labour* were not under the editorship of Mayhew himself. And, indeed, all of Mayhew's work is a product of collaborations with others. As Mayhew notes in the Preface to *London Labour*, Henry Wood contributed so much of the content that he, too, should be considered its author (1:xvi). But this point only underscores the broader point of his reliance on a team of collaborators to complete the project. For a more detailed account of his collaborations, see Schroeder, Leckie, and Herdman ("Working").

19. To be sure, the greatest disturbance to temporality in the mid-nineteenth century—the geological discoveries suggesting that the rock record pre-dated humans and hence contradicted Mosaic law—is neither raised in Eliot's narrator's reference to post-time nor in Mayhew's work. Eliot's narrator's dismay, not to mention his longing for the soporific effect of Sunday morning church services (rather than the jolt that the recent geological discoveries would have delivered), however, seems to indicate a larger disturbance in the field of time, one that would be sealed with the publication of *The Origin of Species* in the same year as *Adam Bede*.

20. Anderson nicely captures the drama and excitement of this period—the sleepless nights, the 24/7 work—in *London Vagabond* (see especially 104–6).

21. See Sheila Smith's excellent study on the blue book and the novel. Novelists also used these reports in similar ways, treating "the novel as though it were a popular form of Blue book in order to make their readers explore social problems" (29). It is worth noting that Mayhew imagines his work as an encyclopedia of London, and this conception also informs his project.

22. See Lesjak and Rosenman for excellent discussions of the Enclosure Acts and their impact on the poor and working classes.

23. See my *Open Houses* for an extended discussion of Parkes's response to the blue book.

24. See Ole Münch and Jenna Herdman.

25. See Schroeder for a detailed account of *London Labour*'s publication history.

26. To be sure, Mayhew's life and work, as many critics have noted, is full of irreconcilable contradictions. His own living conditions were far from settled, and he died bankrupt and alone in the very sort of tenement about which he had so vividly written decades earlier.

27. See Schroeder and Leckie for a discussion of *London Labour*'s form; see Taithe for a discussion of the part edition's wrappings. See Leckie ("Henry") for a discussion of local features of the text.

28. For an excellent study of philology in relation to Mayhew's work, see Taithe (45–59).

29. Notably, the many abridged editions of *London Labour*, seeking greater coherence for Mayhew's unruly work, tame and tie closed its disruptive energies and excise this dimension of his work from the historical record.

30. Carolyn Lesjak makes a similar point when she links the English enclosure of common lands with climate change: "While it may seem a long way from the British countryside to the *chawls* and *favelas* of today's megacities, seeing today's 'urban climacteric' as the logical outcome of the set of practices and policies that coalesced around acts of enclosure in Britain usefully helps us to see the ways in which enclosure was at once a local phenomenon, and one that, from its outset, had global reach, if not the fully realized means to impose its order on a global scale" (17). In some ways, this is the point that thinkers from Heidegger through

Latour make when they argue that the future defines the present and not the other way around.

31. I am indebted to Jesse Oak Taylor for posing this question at his NAVSA seminar in Banff 2018.

32. I have not seen Mosse's installations in a gallery; my access to them has been solely through their print and digital mediations. Mosse points out that the work is intended as a "powerful immersive audio visual installation" (email correspondence); while I have not experienced the installation as Mosse intended, I can attest to how jolting even the short video clips are, and no doubt the gallery experience is even more transformative.

33. This reading departs from Agamben's insofar as it is reading Mosse's interpretation of the camps and arguing that Mosse at once represents the state of exception about which Agamben writes *and* interrupts it to perform or convey interruption itself.

34. "The camera is produced by a multinational defence and security corporation that manufactures cruise missiles, drones, and other technologies. Primarily designed for surveillance, it can also be connected to weapons systems to track and target the enemy" (Mosse, "Transmigration," n.p.).

35. I am grateful to Mosse for his comments on this connection in our email correspondence.

36. "At every point," Franny Nudelman writes, "Mosse aims to slow down the process of producing these images, and to slow the viewer as well in her efforts to consume them" (2). Interestingly, Mosse notes that this slowing down also heightens, in the context of digital technology used in courtrooms, the viewers' sense that they are watching a premeditated crime and that the subject is guilty (n.p.). See also Benjamin's "The Work of Art" essay and its treatment of film as a technology capable of making hitherto unseen movements visible.

37. From Timothy Clark's questioning of what further cultural representations can contribute to climate action to the recognition that further information in general is not required, there is widespread agreement that documentary realism has failed to generate the shift in climate sensibility it so often sought. See Clark and Barbara Herrnstein Smith, among many others.

38. I have used Andrew Benjamin's adapted translation for this citation (cited 111).

39. See Andrew Benjamin's reference to Benjamin's dialectical image as a "temporal montage" (111) that breaks with continuity to privilege simultaneity and enable multiple temporalities to register at once (see also Buck-Morss, 55–56; Ferris, 16).

40. Benjamin, Löwy writes, revises Marx to write "*a Marxism of unpredictability*" (*Fire Alarm*, 109). In this formulation Lowy adds the future to the dialectical image's collapsing of past and present: if "history is open, if 'the new' is possible, this is because the future is not known in advance; the future is not the ineluctable

result of a given historical evolution, the necessary and unpredictable outcome of the 'natural' laws of social transformation, the inevitable fruit of economic, technical or scientific progress" (109).

41. If, as Benjamin argues, linear empty time requires rethinking, so too does the model of representation to which it is linked. Interestingly, Benjamin's posthumous *Arcades Project* approximates this sort of rethinking in a form that resembles both Mayhew's *London Labour* and Mosse's work in general. Eliot, while often interrupting the chronology of her novels, could not quite imagine these new representational models (although her last work, *Theosophratus Such*, comes close). Like Benjamin's *Arcades*, Mayhew's *London Labour* seeks to capture the rise of the city and modernity—in his case London rather than Paris—through found material on diverse topics often left to signify laterally. Like Benjamin's focus on ragpickers, Mayhew's focus on scavengers informs his own approach. And like Benjamin's unfinished *Arcades*, Mayhew's unfinished *Labour* disavows chronology. Mosse's *The Fall*, with its images of found objects of airplanes, also resonates here.

42. "The biggest quarrel he [Mosse] has with documentary photography is its embrace of urgency, and the speed that urgency requires. The photojournalist works against a deadline, and is rewarded for capturing a moment of maximum danger" (Nudelman, 5). See also Robert Smithson, Pierre Huyghe, Christi Belcourt for artists who integrate questions of time and ecology into their art practices.

43. Instead of seeing what the viewer sees, we see him seeing; and instead of seeing directly we see him looking *through binoculars*. This image, itself mediated by Mosse's "slow" media technology, underscores the mediation of everything we see.

44. Jeffrey Jerome Cohen writes that "to lay hand upon stone is to press against time in material form" and to realize "that stone's time is not ours, that the world is not for us" (cited in Menely, "Ecologies," 87).

ALARMING!

1. I would like to thank Michal Leckie for giving me the idea for the formatting of this chapter.

2. To be sure, the increasing frequency of wildfires supports the "house on fire" claim; when they spread to suburbs and towns, houses are, in fact, on fire. Further, Thunberg's analogy performs the thinking she wants to produce: instead of thinking individually about only the house in which we live, she wants us to think collectively about the earth as "our house." The question I am pursuing here is the success of that performative. That is, does it short-circuit as an empty threat because it continues to signify mainly on the individual level or does it incite the response to a real and urgent problem that Thunberg seeks?

3. See the Introduction to Menely and Taylor's *Anthropocene Reading* for a good discussion of the debates with respect to the origin of the Anthropocene. See Choi and Leckie for a discussion of temporality in relation to climate change.

4. From *We Have Never Been Modern* to *Facing Gaia* and beyond, Latour has been writing in the register of climate warnings gone unheeded.

5. The youth movement, communities immediately impacted by climate change, and Black and Indigenous communities advocating for action, on the other hand, have responded forcefully. And more BIPOC communities are worried about climate change than white communities (https://climatecommunication.yale.edu/publications/race-and-climate-change/). Needless to say, though, those communities that are already overburdened with the daily labor of maintaining the lives of their families and themselves do not have extra reserves of time or energy to dedicate to climate justice and action. I'm indebted to Sara Adams for discussion on these points.

6. He later writes: "So it is not as though people haven't been warned, not as though the alarm systems have been angrily unplugged; no, the sirens have been blaring full blast, but a virile decision has nevertheless been made not to let oneself be *inhibited* by the dangers. If there is inhibition, in contrast, it concerns the speed of reaction to catastrophes generated later on. The two attitudes clearly go hand in hand: disinhibition for action where the future is concerned; inhibition when reckoning with retroactive consequences" (*Facing Gaia*, 191).

7. To be sure, both Latour's and Klein's you/we is directed to those who occupy places of privilege in the Global North. They share the call for a broader appreciation of the ways in which "we" are bound up with structures of power that promote and continue "systemic racism/colonialism/racial capitalism that devalues the lives of Black, brown and other oppressed communities" (Sara Adams, personal correspondence). And they similarly share the call for an expanded sense of "our" reliance on and interconnection with nonhuman life (indeed, this is the point Latour makes when he critiques the nature/culture binary). That said, Zoe Todd's critique of Latour in "Indigenizing the Anthropocene" is well taken.

8. This is one of Latour's many examples of the description itself being a prescription: "In practice, the difference between *constative* and *performative* statements (to use the vocabulary of linguists), even though it has been of great concern to philosophers, has always been very slight" (*Facing Gaia*, 47). Most people see the hockey stick graph, for example, and they believe it. They understand that such data "concern us so directly that their mere expression sounds like an alarm" (*Facing Gaia*, 47).

9. Latour invokes the language of panic here to underscore his point. He also implies that panic—indeed a panic attack—will prompt action (*Facing Gaia*, 46) but, as I elaborate below, I'm skeptical of such claims.

10. Latour does follow this point by noting that actions begin with facts that are extended into warnings and point toward decisions.

11. To extend the fire metaphor: our tension throws fuel on that formation, feeds it and helps it to grow. After 9/11 the US government, for example, used fear to subdue political unrest and to ensure that the entire population was more willing to undertake actions to quell that fear. Amber alerts in airports, for example,

put Americans "under tension." And as anyone who has felt such a tension knows, it seeks release. One wants to dial down the tension, to relieve it, and, after a certain point, almost any action will do.

12. See Foer for a discussion of his own climate denial, Norgaard for a discussion of its impact on a Scandinavian community, and MacDuffie ("Charles Darwin") for an excellent prehistory to this structure of denial in the nineteenth century.

13. Although on one occasion Malm does refer to the "fire alarm ringing" in relation to climate change (*Fossil Capital*, 379)

14. He records a conversation from the 1830s in which one of the interlocutors notes that in addition to other detractions, the steam engine had also changed the country's weather, producing summers of "unusual, exampled heat" that prompted "alarm and dread" (*Fossil Capital*, 225).

15. With the advent of the railway, "'[n]atural space,'" Michael Freeman writes, "was replaced by a very different sort of spatiality, one part of which was the straight line" (see also Fyfe [72] and Matheison [17]). As one commentator in the period puts it: "That word line has become a staple of the language" ("Railway Mania 1845," 234).

16. Malm's reading of critical geography studies is ungenerous but his call for an "epoch of diachronicity" (*Fossil Capital*, 8) is well taken *if* it is put into dialogue with other currents (including spatial studies).

17. Latour, for example, writes: "We have become the people *who could have acted* thirty or forty years ago—and who did nothing, or far too little" (*Facing Gaia*, 9, emphasis in original). As I mentioned in "Interruption," that "knowing" itself is by no means straightforward and is better understood as the cognitive dissonance reinforced by temporal frames and conceptual categories that are inadequate to the crisis itself.

18. This a moral conundrum peculiar to the Global North that puts its emphasis on "the person" rather than on the collective's responsibility to future generations. Similarly, as Rob Nixon illustrates, the *slow* unfolding of climate decisions makes it easy to forget that there will be unknown consequences for others based on our actions now.

19. Climate warnings, for example, don't have the clarity of the cigarette package (another not fully successful warning structure oriented toward the future). Cigarette warnings at least tie potential future illness to current actions and issue a clear directive to a single individual: Don't smoke.

20. Leland de la Durantaye writes: "Benjamin's *Theses* identify *catastrophe* that is present and ongoing: our current state of affairs and the model of progress that underlies it. To halt the *catastrophe* we have to interrupt not only this continuum but also the model of time on which it is based" (239).

21. Its formal inventiveness has been oft-noted: it is "a performance; it is an experiment; it is a work permanently in progress" (Marcus, xxv), "a new and radical literary form" (Jennings, "Introduction" 1), and a work with an "elliptical structure" (Huyssen, 149). Several commentators have also noted that, apart from the

opening two entries and the final one, the texts seems to follow no particular order and conform to no plan (Huyssen, 147).

22. The feuilleton, coined as "the little form," is also an example of how "writers shaped their writing practice in order to accommodate the new form" (Jennings, "Walter Benjamin," 258).

23. Comay cites from Benjamin's *Arcades Project* as follows: "That things 'just go on,' writes Benjamin, '*is* the catastrophe. It is not that which lies ahead but that which is always given'" (277).

24. Peter Sinnema argues that the *Illustrated London News* sought to protect readers and soften the images ("Representing"). Paul Fyfe notes similarly how the images cover up problems with industrial modernity.

25. This passage also recalled for me the banner from Cognito's climate emergency camp: "Walk to Waken the Nation."

26. See also Benjamin's discussion of predictive warnings and being too late in relation to fortune tellers in *One-Way Street* (87–89).

27. While I do not situate my discussion in this chapter in the context of Agamben's elaboration of Benjamin's reading of emergency (especially as it relates to Schmitt), his points are clearly relevant to many aspects of the climate crisis. It is here, too, that questions of language come to the fore. For some, like Agamben, the state of emergency is understood as a state of exception. The phrasing, then, directs interpretations in somewhat different directions. This is a debate I am interested in (especially with respect to the "fictional" or performative dimensions of power) but that I am also sidestepping here in the interests of sharpening my understanding of emergency. The climate crisis *is* an emergency, and I want to make the language of emergency signify again. I locate its force, as does Agamben, in Benjamin's call for a *real* emergency. When the alarm reinforces the logic of linear, progressive time, it is vulnerable to being used by the state in all the ways we have seen. See Agamben's *Homo Sacer* and *State of Exception* for an elaboration of his position on these points. See de la Durantaye for a good overview of Agamben's position in the context of Benjamin (81–120).

28. It is interesting, in this context, to consider the rise of climate youth movements around the world. They preempt the state's role in declaring a climate emergency by doing so themselves.

29. The *deliberate* obfuscation, by some corporations, media, and political messaging, of the threat climate change posed should also not be ignored.

30. I am indebted to Sara Adams for stressing this point and for help with this chapter overall.

LAYERING

1. See "Interruption" in this volume.

2. See Roy Scranton's journalistic *Learning to Die in the Anthropocene* and Robert Bringhurst and Jan Zwicky's poetic *Learning to Die: Wisdom in the Age of Climate Crisis*.

3. See also Thesis 13 in Benjamin's "Theses."

4. Macfarlane's extended comments here are quite lovely: "When viewed in deep time, things come alive that seemed inert. New responsibilities declare themselves. A conviviality of being leaps to mind and eye. The world becomes eerily various and vibrant again. Rock has tides. Mountains ebb and flow. Stone pulses. We live on a restless Earth" (*Underland*, 15–16). In an interview he further notes that a "deep time ethic" should have the effect of "charging us with a sense of responsibility *now*, minute by minute, for the legacies we are leaving behind as communities and as polities—and also of pristinating the present moment into greater clarity and wonder. Implausible as it may seem when viewed within epochs and eons our billion-year-old earth, we *do* exist. You do, I do, somehow, now" (Ackerman, 71–72). These comments also resonate with some Indigenous ideas of vibrancy (see Kimmerer).

5. The story of geology is, of course, a story of interruptions—what geologists call *unconformities* and *discontinuities*.

6. Benjamin sees stories as extended into the earth and the "lithic as narrative medium" (Menely and Taylor, 1); as Robert Pogue Harrison notes, even the medium with which we write or print is reducible to earth (14).

7. I am grateful to one of the anonymous readers of this manuscript for suggesting that I elaborate on this point.

8. Tobias Menely and Jesse Taylor Oak's edited collection, *Anthropocene Reading*, was important to me, not least because two of its contributors—Cohen and Mentz—experiment with the form of the academic essay.

9. I include square brackets here and throughout to indicate the places where I have added ellipses not present in the original texts.

IN THE IDIOM OF THE SELF-HELP GUIDE

1. See Zygmunt Bauman, James Surowiecki, and Julia Wright. I am immensely grateful for the suggestions on this chapter offered by one of my anonymous readers.

2. I established these categories from a search of nineteenth-century titles.

3. Peter Sinnema rightly warns against tracking Smiles too closely with the current glut of self-help literature, but there are nevertheless several resonant parallels with respect to procrastination ("Introduction").

4. For an interesting rethinking of case studies in general, see Anja Kanngieser and Zoe Todd's "From Environmental Case Study to Environmental Kin Study."

5. See my "Sequence and Fragment" for a discussion of Smiles's fragmented form in the context of climate change. See both "Sequence and Fragment" and Tina Choi and my article, "Slow Causality," for further parallels between *Self-Help* and *Middlemarch*.

6. Price notes that Eliot's "ruthlessly excerpted" novels were "chopped into anthology-pieces, recycled as calendar decorations, used to test army officers, deployed in a Zionist tract, plastered onto billboards, and quarried for epigraphs" (9).

7. It is worth noting that Eliot also integrated parables, albeit with what might be called a Benjaminian twist, into her novels. For a discussion of this practice in relation to *Middlemarch* see my *Open Houses* (186–94).

8. Since the 1970s, procrastination has received an enormous amount of critical attention from psychologists; more recently, in the last fifteen years or so, philosophers have also joined the discussion. In both contexts, definitions have been central. See, for example, the collection edited by Andreou and White, *The Thief of Time: Philosophical Essays on Procrastination*, as well as Mark Kingwell's essays "Meaning to Get to" and "'We Shall Look Into It Tomorrow.'" I take the OED definition that many psychologists, and others, find useful: "to defer action, especially without a good reason." See Gjelsvik for a good overview of alternative definitions (99–100). Surowiecki writes: "The essence of procrastination lies in not doing what you think you should be doing, a mental contortion that surely accounts for the great psychic toll the habit takes on people. This is the perplexing thing about procrastination: although it seems to involve avoiding unpleasant tasks, indulging in it generally doesn't make people happy" (n.p.).

9. The format of these lessons is distilled from the many, many procrastination books and websites currently in circulation. Some variations on the "lessons" I have outlined in this chapter can be found, in particular, in Neil Fiore's *The Now Habit*, Timothy A. Pychyl's *Solving the Procrastination Puzzle*, and Piers Steel's *The Procrastination Equation*.

10. See Levin et al. for a discussion of some of the pitfalls to thinking in terms of solutions in relation to climate change.

11. Feinberg offers a good overview of Casaubon's goals and the meaning of myth in the period (20–21). Hertz notes that a passage in "Silly Novelists" on the historical imagination with respect to the difficulties of reconstructing "the fragments into a whole" resonates with Casaubon's work (31).

12. Kingwell notes that "reading just one more book or article" is the classic approach of the procrastinating scholar "which helps to explain why there are so many such books and articles being produced, despite much putting off of writing: the general demand for written scholarship's displacing power is sufficient to overcome the particular ineffectiveness of a given procrastinating scholar" ("We Shall Look Into It," 220).

13. Kingwell is realistic here: "To submit to publication is necessarily to embrace the futility of accepting that you cannot say what you mean to say" ("We Shall Look Into It" 6).

14. In his analysis of Casaubon, Hertz focuses on Chapter 20, but he does not discuss the procrastinating structure of the chapter and he ignores the precipitating cause of Dorothea's tears—the argument about procrastination with Casaubon—preferring to focus instead on the broader reasons for her distress: the confusion of Rome, the inadequacy of her education in preparation for it, and the general discord between husband and wife. These points, to be sure, are

all elaborated by the narrator, but it nevertheless makes sense to attend to the precipitating incident as well.

15. Casaubon's work is often figured in terms of weight. See, for example, the following passage that seeks to trace the change in Dorothea's perception: "Besides, had not Dorothea's enthusiasm especially dwelt on the prospect of relieving the weight and perhaps the sadness with which great tasks lie on him who has to achieve them?—And that such weight [his work] pressed on Mr Casaubon was only plainer than before" (135).

16. Here is Welsh's interpretation: "We do not have access to all ordinary human life; there is more life than we could possibly know, let alone respond to; there is more information in general than we can cope with, and if very much of it came at us all at once, we could not physically withstand it" (220). He also compares this passage, aptly, to the introduction of *Felix Holt*.

17. This contrast between an imagined future and the real present also comes up forcefully when Dorothea returns to Lowick after her honeymoon: "The ideas and hopes which were living in her mind when she first saw this room nearly three months before were present now only as memories: she judged them as we judge transient and departed things" (258).

18. Pychyl cites research, for example, that demonstrates that "chronic procrastinators in particular prefer not to have feedback . . . if they have the choice" (40).

19. Geoff Dyer's light-hearted discussion of procrastination takes up this point as follows: "People need to feel that they have been thwarted by *circumstances* from pursing the life which, had they led it, they would not have wanted; . . . Most people don't want what they want: people want to be prevented, restricted. . . . That's why children are so convenient: you have children because you're struggling to get by as an artist . . . or failing to get on with your career. *Then* you can persuade yourself that your children prevented you from this career that had never looked like working out" (129).

20. Haight also notes, aptly, that this passage is suggestive of Casaubon's sexual performance with Dorothea: the red drapery "spreading itself everywhere" in St Peter's is suggestive of Dorothea's sexual experience with Casaubon as "violent and painful" (259–60).

21. D. A. Miller offers a nice summary of this procrastinator's dilemma in terms of Eliot's narrative form in *Middlemarch* (a comment that is itself a pause, placed as it is in parentheses): "(This ambiguity [distinguishing between misrepresentation and truth] will be important in shaping the narrator's own performance, which oddly combines a confidence in his superior version of narrative order with a hesitant, almost sotto voce confession of its ultimate inadequacy)" (143).

22. Haight writes that the "two most important studies of German," in fact, "were already available to him: Creuzer's *Mythologie* (1810–12) had appeared in a

French translation and Lobeck's *Aglaophamus* (1829) was written in Latin" (262). Wiesenfarth, however, writes that Will is correct in his assessment: "Unable to read German, Casaubon is unaware of the work of Karl Otfried Muller, which in 1825 changed the course of studies in mythology" (367).

23. Compare the word "deposited" here to this later reference in which "deposit" aligns with Casaubon's writing in a less flattering light: "But there are some kinds of authorship in which by far the largest result is the uneasy susceptibility accumulated in the consciousness of the author—one knows of the river by a few streaks amid a long-gathered deposit of uncomfortable mud. That was the way with Mr Casaubon's hard intellectual labours" (288).

24. Welsh, for example, writes, "George Eliot makes abundantly clear that virtually the sole motive of Casaubon's scholarship is fame, and that this relation between his knowledge and his being known runs far deeper than mere vanity.... He is subject to the deepest terror by authorship" (238).

25. See Tambling for an excellent discussion of *parergon* in relation both to nineteenth-century usages and to Derrida's reading of this term through Kant (956–58). Following Derrida, the *parergon* is precisely the writing that makes a work impossible to delimit or contain. Not only does Casaubon's "scholarship dissolve from the projected great work into parerga, which lacking the great design, are fragments and witnesses to the fragmentary in their subject-matter, but he himself has to face the dissolution of his own body" (958).

26. For a discussion of the pros and cons of different individual actions, see Wynes and Nicholas, "The Climate Mitigation Gap: Education and Government Recommendations Miss the Most Effective Individual Actions."

27. It should go without saying that such environments are not exempt from criticism. Criticism helps to hone approaches and advance ideas. But it is only helpful when it is issued divorced from the actor's identity and value.

28. I'm aware that Casaubon could also be perceived as articulating a positive procrastination lesson here along the lines of not letting one's self be distracted. And, indeed, as the novel proceeds, he comes closer and closer to modelling good writing practices.

29. Admittedly, they do this work in the middle of the night, which flouts all procrastination advice about getting enough sleep.

30. In recent years, there has been a proliferation of climate action groups that offer a wide range of perspectives and approaches. Many cities have local chapters of 350.org, the Green New Deal (versions of which vary from country to country), Extinction Rebellion, Fridays for Future, and Future Rising, to name only a very few of the many active groups addressing the climate crisis in generative ways. There are also many community groups working for local climate action as well as climate policies for municipal governments. And classes on climate change offer further opportunities for community. Last but not least, if an appealing climate action is not available, then one can start something. Greta Thunberg's movement

began, after all, as one teenager sitting alone on the steps of the Swedish parliament and inviting others to join her.

FOUND QUESTIONS

1. Carson, *Silent Spring*, 10.
2. Kolbert, *Sixth Extinction*, 10.
3. Weston, 188.
4. Franklin, 122.
5. Kimmerer, 174.
6. Scranton, 20.
7. Darwin cited in Lyell, 701.
8. Harrison, 1–2.
9. Heidegger, 5.
10. Weston, 188.
11. Malm, *Fossil Capital*, 3.
12. Klein, *This Changes Everything*, 15.
13. Shelley, "The Triumph of Life."
14. Scranton, 86.
15. Tsing et al., *Arts of Living*, 1.
16. Shotwell, 199.
17. Klein, *This Changes Everything*, 27.
18. Coetzee, 33.
19. Klein, *This Changes Everything*, 27.
20. Weston, 177.
21. Woolf, *A Room*, 5.
22. Greta Thunberg, "School Strike for Climate."
23. Marshall, 1.
24. Williams, "Ideas of Nature," 67.
25. McKibben, 62.
26. John Henry Newman cited in Andrew Miller, 144.
27. Thomas Carlyle cited in Andrew Miller, 145–46.
28. Weisman, 4.
29. Moore, 7.
30. McKibben, 60.
31. Williams, "Ideas of Nature," 68.
32. Ghosh, 119.
33. Kohn, 22.
34. Morgan, "After the Arctic Sublime," 2.
35. Chakrabarty, "Climate," 201.
36. Morgan, "After the Arctic Sublime," 1.
37. NASA, n.p.
38. Latour, *We Have Never Been Modern*, 1.
39. Greta Thunberg, "Speech at the National Assembly in Paris."

40. Barbara Herrnstein Smith, 108.
41. Latour, "Agency," 2.
42. Weston, 181–82.
43. Latour, "Why Has Critique," 227, 230.
44. Lear, "We Will Not be Missed." n.p.
45. Malm, *Fossil Capital*, 394.
46. Malm, *Fossil Capital*, 15.
47. Levinton, 242.
48. Haraway, "Tentacular Thinking," 14.
49. Klein, *This Changes Everything*, 18.
50. Oreskes and Conway, 64.
51. NASA, "Is it Too Late to Prevent Climate Change."
52. Heglar, n.p.
53. Greta Thunberg, "School Strike for Climate."
54. Kolbert, *Sixth Extinction*, 267.
55. Lemenager, *Living Oil*, 23.
56. Greta Thunberg, speech at the National Assembly in Paris, July 23, 2019. https://youtu.be/J1yimNdqhqE
57. Haraway, "Tentacular Thinking," 47.
58. Barbara Herrnstein Smith, 121.
59. Haraway, "Tentacular Thinking," 47.
60. Scranton, 20.
61. Cariou, n.p.
62. Latour, "Agency," 3.
63. Rush, 257.
64. Nixon, 3.
65. Mosse, *Incoming*, n.p.
66. Franklin, 122.
67. Nixon, 14.
68. Coetzee, 30.
69. Watt-Cloutier, xxii.
70. Haraway, "Tentacular Thinking," 35.
71. Lemenager, *Living Oil*, 68–69.
72. Weisman, 5.
73. Lear, n.p.
74. Lyell, 28.
75. Barbara Herrnstein Smith, 115.
76. Nixon, 14–15.
77. Shotwell, 5.
78. Glissant, 206.
79. Glissant, 183.
80. Kimmerer, ix–x.
81. Macfarlane, *Landmarks*, 346.

82. Bachwail cited in Hackett, 13.
83. Todd, 244.
84. Hensley, "After Death," 401.
85. Taylor, *Sky of Our Manufacture*, 17.
86. Haraway, "Tentacular Thinking," 46.
87. Macfarlane, *Landmarks*, 349.
88. Savoy, 43.
89. Chakrabarty, "Climate of History," 216.
90. Nixon, 15.
91. Cariou, n.p.
92. Moore, 6.
93. Pope John Paul, *Laudatio Si'*, 190.
94. Tsing, *Mushroom*, 5.
95. Tsing, *Mushroom*, 254.
96. Baichwal, 202.
97. Nixon, 62.
98. Wynter and McKittrick, 45.
99. Klein, *This Changes Everything*, 28.
100. Ghosh, 8.
101. Carson, *Silent Spring*, 133.
102. Greta Thunberg, "School Strike for Climate."
103. Latour, *We Have Never Been Modern*, 1.
104. McManus, 45.
105. Weisman, 5.
106. Berry, 128.
107. Kolbert, *Sixth Extinction*, 261.
108. Foer, 35.
109. Long Soldier, 43.

FRANKENCLIMATE

1. Scholarly convention now labels Frankenstein's creation, the Creature. In this chapter, however, I follow Shelley's lead by using different appellations at different times in my commentary. The logic I follow in my own designations is as follows: when I deploy *monster* I am generally referencing an idea that escapes categorization or containment, and when I use *Creature* I refer to those places where the Creature exhibits a more contained and coherent subjectivity. But this demarcation does not always hold, for obvious reasons. Unless otherwise indicated, all citations are to the 1818 edition of *Frankenstein*.

2. See Siobhan Carroll for an excellent study of the novel in relation to Arctic exploration and global cooling. For readings of the novel in the context of technology see Thomas H. Ford's "Frankenscription," Smith's "Frankenstein in the Automatic Factory," and Willis's "Scientific Self-Fashioning."

3. Benjamin Morgan alerted my attention to John Luther Adam's palindromic musical piece, "Become Ocean," in the context of deep time. This piece also resonates nicely with Shelley's novel and with efforts to think climate change through alternative temporalities.

4. This is the case for most film adaptations as well as for Andrew Ager's 2019 opera adaptation of the novel.

5. See, for example, Newman and Brooks.

6. See Mary Lowe-Evans for another reading of the letter as contract.

7. Johnson, notably, also wrote one of the most influential essays on Shelley's novel.

8. It is worth noting that Derrida makes this comment in a footnote and that Johnson modifies the translation as follows: "The structure of the framing effects is such that no totalization of the border is even possible. Frames are always framed: thus, by part of their content" (231). The original is as follows: "mais que la structure des effets d'encadrement est telle qu'aucune totalisation de la bordure ne peut même s'en produire. Les cadres sont toujours encadrés: donc par tel morceau de leur contenu (513).

9. Poe's short story famously concludes with Dupin's final trick. He "returns" the letter but does not want to leave the "interior blank" and so, he tells the narrator, "I just copied into the middle of the blank sheet" a quotation from Crébillon (23).

10. See Allen MacDuffie's excellent essay on nineteenth-century versions of climate denial ("Charles Darwin").

11. For a reading of this novel in relation to history that chimes with my treatment of time here, see Freeman, *Time*, 95–105. She develops "the body as a method" and argues that *Frankenstein* "offers us figures for witnessing the history of a discredited form of knowledge and for tracking its afterlife" (96).

12. As Ford puts it, there is "a fundamental restructuring in this moment [1800] of the categories of European knowledge" (274).

13. Some place-names in this chapter are the prior spellings that are used in *Frankenstein* rather than the current names. See, for example, Petersburgh, Chamounix, and Archangel.

14. Walton notes that the Russian "sledge" is "far more agreeable" than the English stagecoach. In doing so, he anticipates mail delivery and, indeed, by the end of the paragraph, he explains that he will be travelling on the "post-road," the very road along which his letter to Margaret will travel, to Archangel.

15. In a strange repetition, Walton both begins and ends this list of potential discoveries with the possibility of discovering "the wondrous power which attracts the needle" (7) and "ascertaining the secret of the magnet" (8), as if his own reveries have lost direction in the recording.

16. Walton begins his expedition before the French and English revolutions that would undermine confidence in Enlightenment rationality and, with it, ideas of progress, but Shelley was writing after these revolutions and in the midst of the

Industrial Revolution with its joint commitment to industrial progress and vivid display of its shortcomings. See Montag.

17. Interestingly, the diary genre informs much climate change literature over the past few decades. See, for example, Octavia Butler's *Parable of the Sower*, Helen Simpson's "Diary of an Interesting Year," Megan Hunter's *The End We Start From*, and Jeff VanderMeer's *Annihilation*, to name only a few very different diary-form narrations.

18. I am indebted to Susan Wolfson's chronology for the events on these dates (xxv).

19. One thinks, too, of Percy Shelley's comment in a letter to Peacock that "rivers are not like roads, the work of the hands of man; they imitate mind, which wanders at will over pathless deserts, and flows through nature's loveliest recesses" (cited in Dowden, 28).

20. Frankenstein says to Walton: "All my speculations and hopes are as nothing, and like the archangel who aspired to omnipotence, I am chained in an eternal hell" (147).

21. The Creature arrives in the novel in another blank defined by the white, borderless landscape: "About two o'clock the mist cleared away, and we beheld, stretched out in every direction, vast and irregular plains of ice, which seemed to have no end" (13).

22. We should pay attention to what happened when Mount Tambora erupted because, as many critics note, the most popular form of geoengineering (or climate engineering) today is "the injection of reflective sulfate aerosol particles into the stratosphere—the upper atmosphere—to mimic the effect of large volcanic eruptions, which have cooled the planet temporally in the past" (Bjornerud, 15). Bjornerud notes that this technique is now called "Solar Radiation Management." (Another example that is often given in this context is the eruption of Mount Pinatubo in the Philippines in 1991.)

23. See especially Carroll.

24. See also Hodges for discussion of novel sequence and feminism. The Creature, she writes, is a "figure of disruption" that is "never fully contained" (158).

25. Consider the rhyming beginnings to these two sections of return, as if they are returning on themselves: "The die is cast; I have consented to return" and "It is past; I am returning to England" (150).

26. For two different readings of this scene see Moretti and Spivak.

27. This comment echoes the Creature's earlier appeal to Frankenstein: "Was I then a monster, a blot upon the earth, from which all men fled, and whom all men disowned?" (81). By extending his criminality to "all human kind" in the passage cited above, the Creature indicts those who flee or disown monstrosity as monsters and criminals themselves. This observation has an obvious parallel in relation to climate change action and the failure of response in the Global North.

28. The De Laceys call the monster their "spirit of good" (141). Frankenstein also imagines the monster as himself: "my own spirit let loose from the grave, and forced to destroy all that was dear to me" (49).

THE ACADEMIC BOOK, INTERRUPTED

1. Of course two readings are possible here: Frankenstein imagines he's saving the world, on the one hand, but he is also foreclosing an alternative form of community, on the other. See Talia Schaffer's *Communities of Care* for a wise and moving elaboration of communities that meet the needs of others.

2. I take this reference to braided narratives from Robin Kimmerer and, while this book has focused on traditions in the Global North spanning the last two centuries, I cannot imagine any approach now that does not turn to the vast range of Indigenous knowledges from which the Global North, and white settlers in particular, still have so much to learn. In this book, I have returned to variations on "thinking otherwise" in writers whose work, in general, has been either folded into prevailing narratives of linearity and progress or marginalized. I very much appreciate those Indigenous thinkers—knowledge-keepers—whose approaches have stayed the course to help all of us navigate the complexities of the climate crisis. Much of the work in this area offers acrobatically multidimensional approaches to time, recognizes and respects the connections and vibrancy of the sensible world in this context, and has never adopted the attitude of the Earth as a resource for human activity, an attitude that has produced the crisis we are all living now (see, in particular, Cariou, Estes, Kimmerer, Tallbear, and White).

3. I am grateful to one of the anonymous readers of this book for this observation.

4. In a lovely passage on Pater, Andrew Miller writes: "Alert to the responsiveness of those about whom he writes, making that responsiveness visible and appealing in his own prose, Pater aims to inspire responsiveness in his own readers. In this way we are granted company in the very act in which we are engaged, our responsiveness anticipated familiarly within the text, by the text" (18–19). James Wood offers something similar but different in his contrast between, as a critic, writing *about* texts and writing *through* them (5–13, 407–25). The latter—writing through (and with)—the text resonates with Miller's comments here.

5. I have thought often of this "we" as I have been writing this book. I began with a comment on Wynter's "referent-we" in a footnote in "Interruption." I'm returning to this question, no more sure than I was at the outset, with a reference to Dionne Brand's reflection on the "we," via Christina Sharpe. Her comments are especially relevant for this book insofar as they focus on nineteenth-century literature as exercising a noninclusive mode of address. She writes: "*We* has a certain barbarity to it—a force. It is an administrative category. . . . To read is to encounter this *we* at every juncture, even when the word is not invoked, even in its most benign well-meaning form. I ingested in those early years of reading the

summons and explusion of *we*. The desire to enter; the impossibility of entering if . . ." (29). This is different, graver ellipsis work than those I have explored in this book. I continue to sit with Brand's ellipses and instead of commenting on them here, I will let them stir and simmer thinking for others as they have, also, for me.

6. See my "Desire Paths" for a different angle on this issue.

7. In relation to Darwin's theories, Gillian Beer makes a comment that is apt for us today: "Most major scientific theories rebuff common sense. They call on evidence that is beyond the reach of our senses and overturn the observable world. They disturb assumed relationships and shift what has been substantial into metaphor. The earth now only *seems* immovable. Such major theories tax, affront, and exhilarate those who first encounter them, although in fifty years or so they will be taken for granted, part of the apparently common-sense set of beliefs which instructs us that the earth revolves around the sun whatever our eyes may suggest" (1).

8. At the time of this writing, global carbon emissions continue to *rise* rather than fall even after a year of reduced energy use during the pandemic and dire warnings from world leaders and countries in the Global North.

9. See MacDuffie's "Charles Darwin" for an excellent treatment of this tension and its relation to climate denial across several nineteenth-century works.

10. To be sure, this is a stretch; those other birds were not always happy to incubate the cuckoo's eggs.

Bibliography

Ackerman, Diane. "An Interview with Robert Macfarlane." *Conjunctions* 37 (2019): 63–77.
Adam, John Luther. *Become Ocean*. Cantaloupe Music, October 30, 2014. Orchestral composition.
Agamben, Giorgio. *Homo Sacer: Sovereign Power and Bare Life*. Translated by Daniel Heller-Roazen, Stanford University Press, 1998.
———. *Infancy and History: On the Destruction of Experience*. Translated by Liz Heron. Verso, 1993.
———. *State of Exception*. Translated by Kevin Attell. University of Chicago Press, 2005.
"Alarming and Fatal Accident on the Great Northern Railway." *Illustrated London News*, July 18, 1846, 35–36.
Anderson, Christopher Gangadin. *London Vagabond: The Life of Henry Mayhew*. Christopher Anderson, 2018.
Andreou, Chrisoula, and Mark D. White, eds. *The Thief of Time: Philosophical Essays on Procrastination*. Oxford University Press, 2010.
"Another Fearful Railway Accident." *Illustrated London News*, September 7, 1861, 235–36.
Armstrong, Isobel. *Victorian Poetry: Poetry, Poetics, and Politics*. Routledge, 1993.
Austin, J. L. *How To Do Things with Words*. Oxford University Press, 1962.
Baichwal, Jennifer. "Our Embedded Signal." In Edward Burtynsky, Jennifer Baichwal, and Nicolas De Pencier, *Anthropocene*, 197–204. Edited by Sophie Hackett, Andrea Kunard, and Urs Stahel. Goose Lane, 2018.

Baucom, Ian. *History 4° Celsius: Search for a Method in the Age of the Anthropocene*. Duke University Press, 2020.
Bauman, Zygmunt. "Modern Adventures of Procrastination." *Parallax* 5, no. 1 (1998): 3–6.
Beck, Ulrich. *The Metamorphosis of the World: How Climate Change is Transforming our Concept of the World*. Polity Press, 2016.
Beer, Gillian. *Darwin's Plots: Evolutionary Narrative in Darwin, George Eliot, and Nineteenth-Century Fiction*. Cambridge University Press, 2010.
Benjamin, Andrew. "Benjamin's Modernity." In *The Cambridge Introduction to Walter Benjamin*, edited by David Ferris, 97–114. Cambridge University Press, 2012.
Benjamin, Walter. *The Arcades Project*. Translated by Howard Eiland and Kevin McLaughlin. Harvard University Press, 1999.
———. "The Author as Producer." In *Reflections*. Translated by Edmund Jephcott, 220–38. Harcourt, 1978.
———. *One-Way Street*, edited by Michael W. Jennings and translated by Edmund Jephcott. Harvard University Press, 2016.
———. *Selected Writings*, 4 vols., edited by Howard Eiland and Michael W. Jennings. Harvard University Press, 2003.
———. "Theses on the Philosophy of History." In *Illuminations*, translated by Harry Zohn, 253–64. Harcourt, 1968.
Berry, Wendell. *Given*. Shoemaker, 2005.
Bjornerud, Marcia. *Timefulness: How Thinking Like a Geologist Can Help Save the World*. Princeton University Press, 2018.
Blair, Sara, Joseph B. Entin, and Franny Nudelman. *Remaking Reality: U.S. Documentary Culture After 1945*. University of North Caroline Press, 2018.
Bourdieu, Pierre. *In Other Words: Essays Towards a Reflexive Sociology*. Translated by Matthew Adamson. Stanford University Press, 1990.
Brand, Dionne. *An Autobiography of the Autobiography of Reading*. University of Alberta Press, 2020.
———. "On Narrative, Reckoning and the Calculus of Living and Dying." *Toronto Star*, July 4, 2020. https://www.thestar.com/entertainment/books/2020/07/04/dionne-brand-on-narrative-reckoning-and-the-calculus-of-living-and-dying.html.
Bringhurst, Robert, and Jan Zwicky. *Learning to Die: Wisdom in the Age of Climate Crisis*. University of Regina Press, 2018.
Brooks, Peter. *Body Work*. Harvard University Press, 1993.
Buckland, Adelene. *Novel Science: Fiction and the Invention of Nineteenth-Century Geology*. University of Chicago Press, 2013.
Buck-Morss, Susan. *The Dialectics of Seeing: Walter Benjamin and the Arcades Project*. MIT Press, 1989.
Burges, Joel, and Amy J. Elias, eds. *Time: A Vocabulary of the Present*. New York University Press, 2016.

Buse, Peter, Ken Hirschkop, Scott McCracken, and Bertrand Taithe. *Benjamin's Arcades: An unGuided Tour.* Manchester University Press, 2005.
Butler, Judith. *Frames of War: When Is Life Grievable?* Penguin, 2016.
———. "Jacques Derrida." *London Review of Books* 26, no. 21 (November 4, 2004). https://www.lrb.co.uk/the-paper/v26/n21/judith-butler/jacques-derrida.
———. *Parting Ways: Jewishness and the Critique of Zionism.* Columbia University Press, 2013.
Callaway, Elizabeth. "A Space for Justice: Messianic Time in the Graphs of Climate Change." *Environmental Humanities* 5, no. 1 (2014): 13–33.
Cameron, Emilie. *Far Off Metal River: Inuit Lands, Settler Stories, and the Making of the Contemporary Arctic.* University of British Columbia Press, 2015.
Caracciolo, Marco. *Slow Narrative and Nonhuman Materialities.* University of Nebraska Press, 2022.
Cariou, Warren. "Tarhands: A Messy Manifesto." *Imaginations* 3, no. 2 (2012). http://imaginations.glendon.yorku.ca/?p=3646.
Carlson, J. A. "Just Friends?: Frankenstein and the Friend to Come." *Nineteenth-Century Contexts* 41, no. 3 (2019): 287–302.
Carroll, Siobhan. "Crusades Against Frost: Frankenstein, Polar Ice, and Climate Change in 1818." *European Romantic Review* 24, no. 2 (2013): 211–230.
Carson, Rachel. *The Sea Around Us.* Oxford University Press, 1989.
———. *Silent Spring.* Houghton, 1962.
Castronovo, Russ. "Facts, Faction, Anachronism." *PMLA* 134, no. 5 (2019): 1143–49.
Chakrabarty, Dipesh. "Climate and Capital: On Conjoined Histories." *Critical Inquiry* 41, no. 1 (Autumn 2014): 1–23.
———. "The Climate of History: Four Theses." *Critical Inquiry* 35, no. 2 (Winter 2009): 197–222.
———. *The Climate of History in a Planetary Age.* University of Chicago Press, 2021.
Choi, Tina Young. *Anonymous Connections: The Body and Narratives of the Social in Victorian Britain.* University of Michigan Press, 2016.
Choi, Tina Young, and Barbara Leckie. "Slow Causality: The Function of Narrative in the Age of the Anthropocene." *Victorian Studies* 60, no. 4 (2018): 565–87.
Chun, Wendy Hui Kyong. "On Hypo-Real Models or Global Climate Change: A Challenge for the Humanities." *Critical Inquiry* 41, no. 3 (2015): 675–703.
Clark, Timothy. *Ecocriticism on the Edge: The Anthropocene as a Threshold Concept.* Bloomsbury, 2015.
"Climate and Covid-19: Converging Crises." *The Lancet,* January 9, 2021. https://www.thelancet.com/journals/lancet/article/PIIS0140-6736(20)32579-4/fulltext.
Coetzee, J. M. *The Lives of Animals.* Princeton University Press, 1999.
Cohen, Jeffrey Jerome. "Anarky." In *Anthropocene Reading: Literary History in Geologic Times,* edited by Tobias Menely and Jesse Oak Taylor, 25–42. Pennsylvania State University Press, 2017.

Comay, Rebecca. "Benjamin's Endgame." In *Walter Benjamin's Philosophy: Destruction and Experience*, edited by Andrew Benjamin and Peter Osborne. Routledge, 1994.
Craps, Stef, and Rick Crownshaw. "Introduction: The Rising Tide of Climate Change Fiction." *Studies in the Novel* 50, no. 1 (2018): 1–8.
Crary, Jonathan. *24/7: Late Capitalism and the Ends of Sleep*. Verso, 2013.
Creaghan, Jason. "Confessions of a Climate Kook." *The Oscar* 48, no 4 (April 2020): 13.
Crutzen, Paul J., and Eugene F. Stoermer. "The Anthropocene." *IGBP Newsletter* 41 (2000): 17–18.
Darwin, Charles. *The Origin of Species by Means of Natural Selection or the Preservation of Favoured Races in the Struggle for Life*. Murray, 1859.
De Quincey, Thomas. "The English Mail Coach." In *Joan of Arc: The English Mail Coach*, edited by J. M. Hart, 46–123. Henry Holt, 1893.
Defoe, Daniel. *Robinson Crusoe*. Olive Garden Books, 2013.
De la Durantaye, Leland. *Giorgio Agamben: A Critical Introduction*. Stanford University Press, 2009.
Derrida, Jacques. "Living On/Border Lines." Translated by James Hulbert. In *Deconstruction and Criticism*, edited by Harold Bloom, 62–142. Continuum, 1979.
———. *The Postcard: From Socrates to Freud and Beyond*. Translated by Alan Bass. University of Chicago Press, 1987.
———. *Specters of Marx: The State of the Debt, the Work of Mourning and the New International*. Translated by Peggy Kamuf. Routledge, 1994.
Diamond, Cora. "The Difficulty of Reality and the Difficulty of Philosophy." *Partial Answers: Literature and the History of Ideas* 1, no 2 (2013): 1–26.
Dickens, Charles. *Bleak House*. Penguin, 2003.
Dowden, Edward. *The Life of Percy Bysshe Shelley*. Kegan Paul, 1896.
Duncan, Ian. "George Eliot and the Science of the Human." In *A Companion to George Eliot*, edited by Amanda Anderson and Harry E. Shaw, 471–85. Wiley 2013.
DuVernay, Ava, dir. *When They See Us*. Netflix, 2019.
Dyer, Geoff. *Out of Sheer Rage: In the Shadow of D. H. Lawrence*. Canongate, 1997.
Eiland, Howard, and Michael W. Jennings. *Walter Benjamin: A Critical Life*. Belknap, 2014.
Eliot, George. *Adam Bede*. Penguin, 1985.
———. *Felix Holt*. Penguin, 1959.
———. *Middlemarch*. Edited by Bert G. Hornback. Norton, 1977.
———. *Selected Essays, Poems and Other Writings*. Edited by A. S. Byatt and Nicholas Warren. Penguin, 1990.
Englander, Nathan. "A Man Set Himself on Fire. We Barely Noticed." *New York Times*, April 20, 2018. https://www.nytimes.com/2018/04/20/opinion/david-buckel-fire-prospect-park.htm.

Enright, Anne. "By the Book." *New York Times*, March 8, 2020, 6.
Erev, Stephanie. "What Is It Like to Become a Bat? Heterogeneities in an Age of Extinction." *Environmental Humanities* 10, no. 1 (May 1, 2018): 129–49. https://doi.org/10.1215/22011919-4385498.
Estes, Nick. *Our History Is the Future: Standing Rock Versus the Dakota Access Pipeline, and the Long Tradition of Indigenous Resistance*. Verso, 2019.
———. "Water Protectors Gave Us a Vision of the Future." *Society + Space* 12 (October 2020). https://www.societyandspace.org/articles/water-protectors-gave-us-a-vision-of-the-future.
Feinberg, Monica L. "Scenes of Marital Life: The *Middlemarch* of Extratextual Reading." *Victorian Newsletter* 77 (1990): 16–26.
Ferris, David. "Introduction: Reading Benjamin." In *The Cambridge Introduction to Walter Benjamin*, edited by David Ferris, 1–17. Cambridge University Press, 2012.
Fiore, Neil. *The Now Habit*. Penguin, 1989.
Foer, Jonathan Safran. *We Are the Weather: Saving the Planet Begins at Breakfast*. Hamish Hamilton, 2019.
Ford, Thomas H. "Frankenscription, a Natural History of Poetry." *Nineteenth-Century Contexts* 41, no. 3 (2019): 271–85.
Foucault, Michel. *The Order of Things: An Archaeology of the Human Sciences*. Vintage, 1973.
François, Anne-Lise. "Shadow-Boxing: Empty Blows, Practice Steps, and Nature's Hold." *Qui Parle* 25, no. 1–2 (2016): 137–77.
———. "Ungiving Time: Reading Lyric by the Light of the Anthropocene." In *Anthropocene Reading: Literary History in Geologic Times*, edited by Tobias Menely and Jesse Oak Taylor, 239–58. Pennsylvania State University Press, 2017.
Franklin, Ursula. *The Real World of Technology*. Anansi, 1999.
Freeman, Elizabeth. "Synchronic / Anachronic." In *Time: A Vocabulary of the Present*, edited by Joel Burges and Amy J. Elias, 129–43. New York University Press, 2016.
———. *Time Binds: Queer Temporalities, Queer Histories*. Duke University Press, 2010.
Freeman, Michael J. *Railways and the Victorian Imagination*. Yale University Press, 1999.
Friedman, Susan Stanford. *Planetary Modernisms: Provocations on Modernity Across Time*. Columbia University Press, 2015.
———. "Scaling Planetarity: *Spacetime* in the New Modernist Studies—Virginia Woolf, H. D., Hilma af Klint, Alicja Kwade, Kathy Jetnil-Kijiner." *Feminist Modernist Studies* 3, no. 2 (2020): 118–47.
Fyfe, Paul. "Illustrating the Accident: Railways and the Catastrophic Picturesque in *The Illustrated London News*." *Research Society for Victorian Periodicals* 46, no. 1 (2013): 61–91.

Gaull, Marilyn. "From the Fossils to the Clones: On Verbal and Visual Narrative." *Wordsworth Circle* 38, no. 1–2 (2007): 77–83.

Ghosh, Amitav. *The Great Derangement: Climate Change and the Unthinkable*. University of Chicago Press, 2016.

Gibson, Lady Jane, ed. *Shelley Memorials: From Authentic Sources*. Wentworth, 1899.

Gidal, Eric. *Ossianic Unconformities: Bardic Poetry in the Industrial Age*. University of Virginia Press, 2015.

Gjelsvik, Olav. "Prudence, Procrastination, and Rationality." In *The Thief of Time: Philosophical Essays on Procrastination*, edited by Chrisoula Andreou and Mark D. White, 99–114. Oxford University Press, 2010.

Glissant, Edouard. *Poetics of Relation*. Translated by Betsy Wing. University of Michigan Press, 1997.

Gould, Stephen Jay. *Time's Arrow, Time's Cycle: Myth and Metaphor in the Discovery of Geological Time*. Harvard University Press, 1987.

Griffiths, Devin. *The Age of Analogy: Science and Literature Between the Darwins*. Johns Hopkins University Press, 2019.

Groom, Amelia, ed. *Time: Documents of Contemporary Art*. MIT Press, 2013.

Grosz, Elizabeth. *The Nick of Time: Politics, Evolution, and the Untimely*. Duke University Press, 2004.

Groves, Jason. *The Geological Unconscious: German Literature and the Mineral Imaginary*. Fordham University Press, 2020.

Guillory, John. "Genesis of the Media Concept." *Critical Inquiry* 36, no. 2 (2010): 321–62.

Hackett, Sophie. "Far and Near: New Views of the Anthropocene." In Edward Burtynsky, Jennifer Baichwal, and Nicolas De Pencier, *Anthropocene*, 13–34. Edited by Sophie Hackett, Andrea Kunard, and Urs Stahel. Goose Lane, 2018.

Haight, Gordon S. "Poor Mr. Casaubon." *Nineteenth-Century Literary Perspectives: Essays in Honor of Lionel Stevenson*, edited by Clyde de L. Ryals, 255–70. Duke University Press, 1974.

Haraway, Donna J. "Anthropocene, Capitalocene, Plantationocene, Chthulucene: Making Kin." *Environmental Humanities* 6, no. 1 (May 1, 2015): 159–65. https://doi.org/10.1215/22011919-3615934.

———. *Staying with the Trouble: Making Kin in the Chthulucene*. Duke University Press, 2016.

———. "Tentacular Thinking: Anthropocene, Capitalocene, Chthulucene." *e-flux Journal* 75 (September 2016). https://www.e-flux.com/journal/75/67125/tentacular-thinking-anthropocene-capitalocene-chthulucene/.

Harrison, Robert Pogue. *The Dominion of the Dead*. University of Chicago Press, 2003.

Hawking, Stephen. *A Brief History of Time*. Bantam, 1988.

Hayot, Eric. "Academic Writing, I Love You. I Really Do." *Critical Inquiry* 41, no. 1 (Autumn 2014): 53–77.

Heidegger, Martin. *The Question Concerning Technology and Other Essays*. Translated by William Lovitt. Garland, 1977.
Heise, Ursula K. *Imagining Extinction: The Cultural Meaning of Endangered Species*. University of Chicago Press, 2014.
Heglar, Mary. "Climate Change Ain't the First Existential Threat." *Resilience*, March 6, 2019. https://www.resilience.org/stories/2019-03-06/climate-change-aint-the-first-existential-threat/.
Hensley, Nathan K. "After Death: Christina Rossetti's Timescales of Catastrophe." *Nineteenth-Century Contexts* 38, no. 5 (2016): 399–415.
———. "Database and the Future Anterior: Reading *The Mill on the Floss* Backwards." *Genre* 50, no. 1 (2017): 117–37.
Hensley, Nathan K., and Philip Steer, eds. *Ecological Form: System and Aesthetics in the Age of Empire*. Fordham University Press, 2019.
Herbert, Christopher. "Rat Worship and Taboo in Mayhew's London." *Representations* 23 (1988): 1–24.
Herdman, Jenna M. "Curious Conversations: Henry Mayhew and the Street-Sellers of *London Labour and the London Poor*." *Journal of Victorian Culture* 26, no. 3 (July 2021): 384–403.
Hertz, Neil. *George Eliot's Pulse*. Stanford University Press, 2003.
Hodges, Devon. "*Frankenstein* and the Feminine Subversion of the Novel." *Tulsa Studies in Women's Literature* 2, no. 2 (Fall 1983): 155–64.
Hoy, David Couzens. *The Times of Our Lives: A Critical History of Temporality*. MIT Press, 2009.
Humpherys, Anne. *Travels into the Poor Man's Country: The Work of Henry Mayhew*. University of Georgia Press, 1977.
Huyssen, Andreas. *Miniature Metropolis: Literature in an Age of Photography and Film*. Harvard University Press, 2015.
Hyde, Lewis. *A Primer for Forgetting: Getting Past the Past*. Farrar, Straus, and Giroux, 2019.
Intergovernmental Panel on Climate Change. "Climate Change Widespread, Rapid, and Intensifying." August 9, 2021. https://www.ipcc.ch/2021/08/09/ar6-wg1-20210809-pr/.
James, Henry. *Literary Criticism: French Writers, Other European Writers, The Prefaces to the New York Edition*, edited by Leon Edel. Library of America, 1984.
Jameson, Frederic. "The End of Temporality." *Critical Inquiry* 29, no. 4 (2003): 695–718.
Jennings, Michael. "Introduction." In *One-Way Street*, edited by Michael W. Jennings and translated by Edmund Jephcott, 1–20. Harvard University Press, 2016.
———. "Walter Benjamin and the European Avant-Garde." In *The Cambridge Introduction to Walter Benjamin*, edited by David Ferris, 18–34. Cambridge University Press, 2012.

Johnson, Barbara. "The Frame of Reference: Poe, Lacan, Derrida." In *The Purloined Poe: Lacan, Derrida, and Psychoanalytic Reading*, edited by John P. Muller and William J. Richardson, 213–51. Johns Hopkins University Press, 1988.

Joshi, Priti. "The Other Great Exhibition: Mayhew's Catalog of the Industrious." *Literature Compass* 9, no. 1 (2012): 95–105.

Kanngieser, Anja, and Zoe Todd. "From Environmental Case Study to Environmental Kin Study." *History and Theory* 59, no. 3 (2020): 385–93.

Kern, Stephen. *The Culture of Time and Space, 1880–1918*. Harvard University Press, 1983.

Kimmerer, Robin. *Braiding Sweetgrass: Indigenous Wisdom, Scientific Knowledge and the Teachings of Plants*. Milkweed Editions, 2013.

Kingwell, Mark. "Meaning to Get to: Procrastination and the Art of Life." *Queen's Quarterly* 109, no. 3, (Fall 2002): 363–81.

———. "'We Shall Look Into It Tomorrow': Kierkegaard and the Art of Procrastination." *Toronto Journal of Theology* 29, no. 2 (2013): 211–26.

Klein, Naomi. *On Fire: The Burning Case for a Green New Deal*. Knopf, 2019.

———. *This Changes Everything: Capitalism vs. The Climate*. Simon and Schuster, 2014.

Klingan, Katrin, Ashkan Sepahvand, Christoph Rosol, and Bernd M. Scherer. *Textures of the Anthropocene: Grain Vapor Ray*. MIT Press, 2015.

Kohn, Eduardo. *How Forests Think: Toward an Anthropology Beyond the Human*. University of California Press, 2013.

Kolbert, Elizabeth. "Now You See It." *New Yorker*, October 15, 2018, 97–99.

———. *Sixth Extinction: An Unnatural History*. Macmillan, 2014.

Krenak, Ailton. *Ideas to Postpone the End of the World*. Translated by Anthony Doyle. Anansi, 2020.

Latour, Bruno. "Agency at the Time of the Anthropocene." *New Literary History* 45 (2014): 1–18.

———. "An Attempt at a 'Compositionist Manifesto.'" *New Literary History* 41, no. 3 (2010): 471–90.

———. *Facing Gaia: Eight Lectures on the New Climactic Regime*. Polity, 2017.

———. *We Have Never Been Modern*. Translated by Catherine Porter. Harvard University Press, 1993.

———. "Why Has Critique Run Out of Steam? From Matters of Fact to Matters of Concern." *Critical Inquiry* 30 (Winter 2004): 225–48.

Lear, Jonathan. "We Will Not Be Missed!" *The Point Magazine*, March 16, 2021. https://thepointmag.com/examined-life/we-will-not-be-missed/.

Leckie, Barbara. "Desire Paths: Nineteenth Century Studies . . ." *Nineteenth-Century Contexts* 43, no. 5 (2021): 581–87.

———. "Henry Mayhew: Urban Ecologist." *Victorian Literature and Culture* 48, no. 1 (Spring 2020): 219–41.

———. "Introduction." In *End-of-Century Assessments and New Directions*. Vol. 6 of *Sanitary Reform in Victorian Britain*, edited by Michelle Allen-Emerson, ix–xxxiii. Pickering & Chatto, 2013.

———. *Open Houses: The Architectural Idea, the Rise of the Novel, and Nineteenth-Century Modernity*. University of Pennsylvania Press, 2017.

———. "Sequence and Fragment, History and Thesis: Samuel Smiles's *Self-Help*, Social Change, and Climate Change." *Nineteenth-Century Contexts* 38, no. 5 (2016): 305–17.

Leckie, Barbara, and Joel Westheimer. "House on Fire." Podcast in six parts. houseonefirepodcast.org.

Lee, Janice. "An Interview with Robin Wall Kimmerer." *The Believer* 134, no. 3 (November 2020). https://believermag.com/an-interview-with-robin-wall-kimmerer/.

Lemenager, Stephanie. *Living Oil: Petroleum Culture in the American Century*. Oxford University Press, 2014.

———. "Sediment." In *Veer Ecology: A Companion for Environmental Thinking*, edited by Jeffrey Jerome Cohen and Lowell Duckert, 168–82. University of Minnesota Press, 2017.

Lerner, Ben. *10:04*. Penguin, 2014.

Lerner, Ben, and Alexander Kluge. *The Snows of Venice*. Spector Books, 2018.

Lesjak, Carolyn. *The Afterlife of Enclosure: British Realism, Character, and the Commons*. Stanford University Press, 2021.

Levin, Kelly, Benjamin Cashore, Steven Bernstein, and Graeme Auld. "Overcoming the Tragedy of Super Wicked Problems: Constraining our Future Selves to Ameliorate Global Climate Change." *Policy Sciences* 45, no. 2 (June 2012): 123–52.

Levine, Caroline. *Forms: Whole, Rhythm, Hierarchy, Network*. Princeton University Press, 2015.

———. *The Serious Pleasures of Suspense: Victorian Realism and Narrative Doubt*. University of Virginia Press, 2003.

Levine, George. *The Realist Imagination: English Fiction from Frankenstein to Lady Chatterley*. University of Chicago Press, 1988.

Levinton, Jeffrey. Afterword to *The Sea Around Us*, by Rachel Carson, 213–43. Oxford University Press, 1989.

Loftus, Donna. "Work, Poverty and Modernity in Mayhew's London." *Journal of Victorian Culture* 19, no. 4 (2014): 507–19.

Long Soldier, Layli. *Whereas*. Graywolf, 2017.

Lovitt, William. Introduction to *The Question Concerning Technology and Other Essays*, by Martin Heidegger, xii–xxix. Translated by William Lovitt. Garland, 1977.

Lowe-Evans, Mary. "Reading with a 'Nicer Eye': Responding to *Frankenstein*." In *Frankenstein*, by Mary Shelley, 215–29. Edited by Johanna M. Smith. Bedford, 1992.

Löwy, Michael. *Fire Alarm: Reading Walter Benjamin's "On the Concept of History."* Translated by Chris Turner. Verso, 2016.

———. "The Revolution Is the Emergency Brake: Walter Benjamin's Political-Ecological Currency." http://web.archive.org/web/20170324221856/http://www.walterbenjaminportbou.cat/sites/all/files/2010_Loewy_ANG.pdf .

Lyell, Charles. *Principles of Geology. Project Gutenberg E-book*, July 22, 2010. https://www.gutenberg.org/files/33224/33224-h/33224-h.htm.

MacDuffie, Allen. "Charles Darwin and the Victorian Pre-History of Climate Denial." *Victorian Studies* 16, no. 4 (2018): 543–64.

———. *Victorian Literature, Energy, and the Ecological Imagination*. Cambridge University Press, 2016.

Macfarlane, Robert. *Landmarks*. Penguin, 2016.

———. *Underland: A Deep Time Journey*. Penguin, 2019.

MacGuire, Richard. *Here*. Pantheon, 2014.

Magnason, Andri Snaer. *On Time and Water*. Translated by Lytton Smith. Biblioasis, 2021.

Malm, Andreas. *Corona, Climate, Chronic Emergency: War Communism in the Twenty-First Century*. Verso, 2020.

———. *Fossil Capital: The Rise of Steam Power and the Roots of Global Warming*. Verso, 2016.

———. *The Progress of This Storm: Nature and Society in a Warming World*. Verso, 2018.

Marcus, Greil. "Preface." In *One-Way Street*, edited by Michael W. Jennings and translated by Edmund Jephcott, ix–xxv. Harvard University Press, 2016.

Marshall, George. *Don't Even Think About It: Why Our Brains Are Wired to Ignore Climate Change*. Bloomsbury, 2014.

Marx, Karl. *Capital*. 3 vols. Translated by Samuel Moore and Edward Aveling. International, 1967.

———. "Theses on Feuerbach." An Appendix to Engels' *Ludwig Feuerbach and the End of Classical German Philosophy*. Progress, 1946.

Mathieson, Charlotte. *Mobility in the Victorian Novel: Placing the Nation*. Palgrave, 2015.

Mayhew, Henry. "Answers to Correspondents." In *The Essential Mayhew*, edited by Bertrand Taithe, 85–251. Rivers Oram, 1996.

———. *London Labour and the London Poor*. 4 vols. Dover, 1968.

———. *London Labour and the London Poor: Selections*. Edited by Janice Schroeder and Barbara Leckie. Broadview, 2020.

Maynard, Robyn. "Against the Carceral State: Making (Black) Freedom in a Time of Crisis and Revolt." Shirley Greenberg Annual Lecture, University of Ottawa, October 8, 2020.

McCarthy, Cormac. *The Road*. Vintage International, 2007.

McKibben, Bill. *End of Nature*. Random House, 2006.

McKittrick, Katherine. *Sylvia Wynter: On Being Human as Praxis*. Duke University Press, 2015.

McManus, Karla. "How Anthropo-scenic!: Concerns and Debates about the Age of the Human." In Edward Burtynsky, Jennifer Baichwal, and Nicolas De Pencier, *Anthropocene*, 45–58. Edited by Sophie Hackett, Andrea Kunard, and Urs Stahel. Goose Lane, 2018.

Menely, Tobias. "Anthropocene Air." *Minnesota Review* 81 (2014): 92–101.

———. *Climate and the Making of Worlds: Toward a Geohistorical Poetics*. University of Chicago Press, 2021.

———. "Ecologies of Time." In *Time and Literature*, edited by Thomas Allen, 85–102. Cambridge University Press, 2017.

Menely, Tobias, and Jesse Oak Taylor, eds. *Anthropocene Reading: Literary History in Geologic Times*. Pennsylvania State University Press, 2017.

Menke, Richard. *Telegraphic Realism: Victorian Fiction and Other Information Systems*. Stanford University Press, 2007.

Mentz, Steve. "Enter Anthropocene, Circa 1610." In *Anthropocene Reading: Literary History in Geologic Times*, edited by Tobias Menely and Jesse Oak Taylor, 43–58. Pennsylvania State University Press, 2017.

Mertens, Mahlu, and Stef Craps. "Contemporary Fiction vs. the Challenge of Imagining the Timescale of Climate Change." *Studies in the Novel* 50, no. 1 (2018): 134–53.

Michaud, Hans. "December 2009, In Conversation with Richard Mosse." *Noah Becker's Whitehot Magazine of Contemporary Art*, December 2009. https://whitehotmagazine.com/articles/in-conversation-with-richard-mosse/1981.

Miller, Andrew. *The Burdens of Perfection: On Ethics and Reading in Nineteenth-Century British Literature*. Cornell University Press, 2010.

Miller, D. A. *Narrative and Its Discontents: Problems of Closure in the Traditional Novel*. Princeton University Press, 1981.

Miller, Elizabeth Carolyn. *Extraction Ecologies and the Literature of the Long Exhaustion*. Princeton University Press, 2021.

Montag, Warren. "'The Workshop of Filthy Creation': A Marxist Reading of *Frankenstein*." In *Mary Shelley's Frankenstein*, edited by Johanna M. Smith, 469–80. Bedford, 2016.

Moore, Jason W. *Capitalism in the Web of Life: Ecology and the Accumulation of Capital*. Verso, 2015.

Moretti, Franco. *Signs Taken for Wonders: On the Sociology of Literary Forms*. Verso, 2005.

Morgan, Benjamin. "After the Arctic Sublime." *New Literary History* 47, no. 1 (2016): 1–26.

———. "*Fin du Globe*: On Decadent Planets." *Victorian Studies* 85, no. 4 (2016): 609–35.

Morton, Timothy. *Ecology without Nature: Rethinking Environmental Aesthetics*. Harvard University Press, 2007.

———. *Hyperobjects: Philosophy and Ecology after the End of the World*. University of Minneapolis Press, 2013.

Morton, Timothy, ed. *A Routledge Literary Sourcebook on Mary Shelley's* Frankenstein. Routledge, 2002.

Mosse, Richard. "Artist Statement." Video sent via private correspondence.

———. *Incoming / Heat Maps*. http://www.richardmosse.com/projects/incoming. 2017.

———. *Incoming. MACK*, 2017.

———. "Through a Glass Brightly: Eastern Congo by Infared." *Pulitzer Center*, November 10, 2011. https://pulitzercenter.org/reporting/through-glass-brightly-eastern-congo-infrared.

———. "Transmigration of the Souls." In *Incoming. MACK*, 2017, n.p.

Muller, John P., and William J. Richardson. *The Purloined Poe: Lacan, Derrida, and Psychoanalytic Reading*. Johns Hopkins University Press, 1988.

Mullen, Mary L. *Novel Institutions: Anachronism, Irish Novels and Nineteenth-Century Realism*. Oxford University Press, 2019.

Münch, Ole. "Henry Mayhew and the Street Traders of Victorian London—A Cultural Exchange with Material Consequences." *London Journal* 43, no. 1 (2018): 53–71.

NASA. "Is It Too Late to Prevent Climate Change?" *NASA Global Climate Change*. n.d. https://climate.nasa.gov/faq/16/is-it-too-late-to-prevent-climate-change/.

Newman, Beth. "Narratives of Seduction and the Seductions of Narrative: The Frame Structure of *Frankenstein*." *English Literary History* 53 (1986): 141–61.

Nichols, Bill. *Representing Reality: Issues and Concepts in Documentary*. Indiana University Press, 1991.

Nixon, Rob. *Slow Violence and the Environmentalism of the Poor*. Harvard University Press, 2011.

Norgaard, Kari Marie. *Living in Denial: Climate Change, Emotions, and Everyday Life*. MIT Press, 2011.

Nudelman, Franny. "Flying Blind." Unpublished conference paper, 2018.

Oreskes, Naomi, and Erik M. Conway. *The Collapse of Western Civilization: A View from the Future*. Columbia University Press, 2014.

Odell, Jenny. *How to Do Nothing: Resisting the Attention Economy*. Penguin, 2019.

Parkes, Bessie Raynor. "The Ladies' Sanitary Association." *English Woman's Journal* 3 (April 1859): 73–85.

Pemberton, Neil. "The Rat-Catcher's Prank: Interspecies Cunningness and Scavenging in Henry Mayhew's London." *Journal of Victorian Culture* 19, no. 4 (2014): 520–35.

Pinkus, Karen. *Fuel: A Speculative Dictionary*. University of Minnesota Press, 2016.

Poe, Edgar Allan. "The Purloined Letter." In *The Purloined Poe: Lacan, Derrida, and Psychoanalytic Reading*, edited by John P. Muller and William J. Richardson, 3–27. Johns Hopkins University Press, 1988.
Pope Francis. *Laudatio Si': On Care for our Common Home*. Libreria Editrice Vaticana, 2015.
Price, Leah. *The Anthology and the Rise of the Novel*. Cambridge University Press, 2000.
———. *How to Do Things with Books in Victorian Britain*. Princeton University Press, 2012.
Pychyl, Timothy A. *Solving the Procrastination Puzzle*. Penguin, 2013.
"Railway Mania 1845." *Illustrated London News* 180 (11 October 1845): 234.
Richardson, Eugene T. *Epidemic Illusions: On the Coloniality of Public Health*. MIT Press, 2020.
Ronda, Margaret. "Mourning and Melancholia in the Anthropocene." *Post45*, June 10, 20136. http://post45.research.yale.edu/2013/06/mourning-and-melancholia-in-the-anthropocene/
Rosenberg, Daniel. "The Trouble with Timelines." In *Time: Documents of Contemporary Art*, edited by Amelia Groom, 60–62. MIT Press, 2013.
Rosenberg, John R. "Introduction to the Dover Edition." In *London Labour and the London Poor*, by Henry Mayhew, v–ix. Dover, 1968.
Rosenman, Ellen. "On Enclosure Acts and the Commons." *BRANCH: Britain, Representation and Nineteenth-Century History*, December 2012. http://www.branchcollective.org/?ps_articles=ellen-rosenman-on-enclosure-acts-and-the-commons.
Rush, Elizabeth. *Rising: Dispatches from the New American Shore*. Milkweed, 2019.
Sachs, Jonathan. "Eighteenth-Century Slow Time: Seven Propositions." *The Eighteenth Century* 60, no. 2 (2019): 185–205.
———. "Slow Time." *PMLA* 134, no. 2 (2019): 315–31.
Said, Edward. *Orientalism*. Vintage, 1978.
Samalin, Zachary. "Plumbing the Depths, Scouring the Surface: Henry Mayhew's Scavenger Hermeneutics." *New Literary History* 48, no. 2 (2017): 387–410.
Savoy, Lauret. *Trace: Memory, History, Race, and the American Landscape*. Counterpoint, 2015.
Schaffer, Talia. *Communities of Care: The Social Ethics of Victorian Fiction*. Princeton University Press, 2021.
Schroeder, Janice. "The Publishing History of Henry Mayhew's *London Labour and the London Poor*." *BRANCH: Britain, Representation and Nineteenth-Century History*, May 2019. http://www.branchcollective.org/?ps_articles=janice-schroeder-the-publishing-history-of-henry-mayhews-london-labour-and-the-london-poor.
Schroeder, Janice, and Barbara Leckie. Introduction to *London Labour and the London Poor: Selections*, by Henry Mayhew, 11–35. Edited by Janice Schroeder and Barbara Leckie. Broadview, 2019.

Schroeder, Janice, Barbara Leckie, and Jenna Herdman. "Working with Mayhew: Collaboration and Historical Empathy in Precarious Times." In *Victorian Culture and Experiential Learning Historical Encounters in the Classroom*, edited by Kevin A. Morrison, 49–63. Palgrave, 2022.

Scott, Heidi C. M. *Chaos and Cosmos: Literary Roots of Modern Ecology in the British Nineteenth Century*. Pennsylvania State University Press, 2014.

Scott, Sir Walter. "From *Edinburgh Magazine* (March 1818)." In *Frankenstein*, by Mary Shelley, 191–96. Edited by J. Paul Hunter, 2nd ed. Norton, 2011.

Scranton, Roy. *Learning to Die in the Anthropocene: Reflections on the End of a Civilization*. City Lights, 2015.

Secord, James. Introduction to *Principles of Geology*, by Charles Lyell, ix–xliii. Penguin, 1997.

Sedgwick, Eve Kosofsky. *Tendencies*. Duke University Press, 1993.

Shakespeare, William. *Henry VI*. In *The Riverside Shakespeare*. Houghton, 1974.

Sharpe, Christina. *In the Wake: On Blackness and Being*. Duke University Press, 2016.

Shelley, Mary. "Editor's Note on the Poems of 1822." In *The Poetical Works of Percy Bysshe Shelley*, edited by Mrs. Shelley, 322–25. Edward Moxon, 1859.

———. *Frankenstein*, edited by J. Paul Hunter, 2nd ed. Norton, 2011.

———. "Introduction." In *Frankenstein*, edited by J. Paul Hunter, 2nd ed., 169–72, Norton, 2011.

Shelley, Percy. "A Defence of Poetry." https://www.poetryfoundation.org/articles/69388/a-defence-of-poetry

———. "The Triumph of Life." Representative Poetry Online, n.d. https://rpo.library.utoronto.ca/poems/triumph-life.

Shotwell, Alexis. *Against Purity: Living Ethically in Compromised Times*. University of Minnesota Press, 2016.

Sinnema, Peter. "Introduction." In *Self-Help: With Illustrations of Character, Conduct, and Perseverance*, vii–xxviii. Oxford University Press, 2002.

———. "Representing the Railway: Train Accidents and Trauma in the *Illustrated London News*." *Victorian Periodicals Review* 31, no. 2 (Summer 1998): 142–68.

Siskin, Clifford, and William Warner. *This Is Enlightenment*. University of Chicago Press, 2010.

Sloterjik, Peter, and Hans-Jürgen Heinrichs. *Neither Sun nor Death*. Translated by Steve Corcoran. Semiotext(e), 2011.

Smiles, Samuel. *Self-Help: With Illustrations of Character, Conduct, and Perseverance. 1859*. Oxford University Press, 2002.

Smith, Barbara Herrnstein. *Practicing Relativism in the Anthropocene On Science, Belief, and the Humanities*. Open Humanities, 2018.

Smith, Russell. "Frankenstein in the Automatic Factory." *Nineteenth-Century Contexts* 41, no. 3 (2019): 303–19.

Smith, Sheila. "Blue Books and Victorian Novelists." *Review of English Studies* 21 (1970): 23–40.

Smithson, Robert. "A Sedimentation of the Mind: Earth Projects." In *Textures of the Anthropocene: Grain Vapor Ray*, edited by Katrin Klingan, Ashkan Sepahvand, Christoph Rosol, and Bernd M. Scherer, 13–32. MIT Press, 2015.
Soldier, Layli Long. *Whereas: Poems*. Graywolf, 2017.
Sörlin, Sverker. "Uncovering the Non-Site: Robert Smithson on Art, Layers, and Time." In *Textures of the Anthropocene: Grain Vapor Ray*, edited by Katrin Klingan, Ashkan Sepahvand, Christoph Rosol, and Bernd M. Scherer, 32–44. MIT Press, 2015.
Spivak, Gayatri. *A Critique of Postcolonial Reason: Toward a History of the Vanishing Present*. Harvard University Press, 1999.
Steedman, Carolyn. "Mayhew: On Reading, About Writing." *Journal of Victorian Culture* 19, no. 4 (2014): 550–61.
Steel, Piers. *The Procrastination Equation: How to Stop Putting Things Off and Start Getting Stuff Done*. Vintage, 2010.
Stull, Kali, and Etienne Turpin, "Our Vectors, Ourselves." *e-flux architecture*, January 2017. http://www.e-flux.com/architecture/superhumanity/68665/our-vectors-ourselves/.
Surowiecki, James. "Later: What Does Procrastination Tell Us about Ourselves?" *The New Yorker*, October 11, 2010. https://www.newyorker.com/magazine/2010/10/11/later.
Szeman, Imre, and Dominic Boyer. "Introduction: On the Energy Humanities." In *Energy Humanities: An Anthology*, edited by Imre Szeman and Dominic Boyer, 1–14. Johns Hopkins University Press, 2017.
Taithe, Bertrand, ed. *The Essential Mayhew*. Rivers Oram, 1996.
Tallbear, Kim. "A Sharpening of the Already Present: An Indigenous Materialist Reading of Settler Apocalypse 2020." Talk presented at the "Humanities on the Brink: Environment, Energy, Emergency" online conference, July 2020.
Tambling, Jeremy. "*Middlemarch*, Realism, and the Birth of the Clinic." *ELH* 57, no. 4 (Winter 1990): 939–60.
"The Tay Bridge Disaster." *Illustrated London News*, January 10, 1880. https://books.google.com/books?id=QYc-AQAAMAAJ.
Tay Bridge Disaster: Report of the Court of Inquiry and Report of Mr Rothery Upon the Circumstances Attending the Fall of a Portion of the Tay Bridge on the 28th December 1879. London, 1880.
Taylor, Astra. *Democracy May Not Exist, but We'll Miss It When It's Gone*. Metropolitan, 2019.
Taylor, Isaac. *Words and Places: or, Etymological Illustrations of History, Ethnology and Geography*. Macmillan, 1865.
Taylor, Jesse Oak. *The Sky of Our Manufacture: The London Fog in British Fiction from Dickens to Woolf*. University of Virginia Press, 2016.
———. "Where Is Victorian Ecocriticism?" *Victorian Literature and Culture* 43 (2015): 877–894.

Thompson, E. P. "The Political Education of Henry Mayhew." *Victorian Studies* 11, no. 1 (1967): 41–62.

———. "Time, Work-Discipline, and Industrial Capitalism." *Past and Present* 38, no. 1 (1967): 56–97.

Thompson, E. P., and Eileen Yeo, eds. *The Unknown Mayhew: Selections from the Morning Chronicle 1849–50*. Penguin, 1973.

Thorsheim, Peter. *Inventing Pollution: Coal, Smoke, and Culture in Britain since 1800*. Ohio University Press, 2006.

Thunberg, Greta. "Speech at the National Assembly in Paris, 23 July 2019." https://youtu.be/J1yimNdqhqE.

———. "School Strike for Climate—Save the World by Changing the Rules," speech at TEDxStockholm, December 12, 2018. https://www.youtube.com/watch?v=EAmmUIEsN9A .

Tiedemann, Rolf. "Dialectics at a Standstill: Approaches to the *Passagen-Werk*." In *The Arcades Project*, by Walter Benjamin, 929–45. Translated by Howard Eiland and Kevin McLaughlin, Belknap, 1999.

Todd, Zoe. "Indigenizing the Anthropocene." In *Art in the Anthropocene: Encounters Among Aesthetics, Politics, Environments and Epistemologies*, edited by Heather Davis and Etienne Turpin, 241–54. Open Humanities, 2015.

Trelawny, E. J. "Shelley's Last Days." *Athaneaum*, no. 2649 (August 3, 1878): 144.

Trexler, Alan. *Anthropocene Fictions: The Novel in a Time of Climate Change*. University of Virginia Press, 2015.

Tricker, Cecilia. "Interview with Mary Ruefle." *The White Review*, no. 24 (March 2019). https://www.thewhitereview.org/feature/interview-mary-ruefle/.

Tsing, Anna Lowenhaupt. *The Mushroom at the End of the World: On the Possibility of Life in Capitalist Ruins*. Princeton University Press, 2017.

Tsing, Anna L., Jennifer Deger, Alder Keleman Saxena, and Feifei Zhou, eds. *Feral Atlas: The More-Than-Human Anthropocene*. Stanford University Press, 2021. http://doi.org/10.21627/2020fa.

Tsing, Anna Lowenhaupt, Heather Anne Swanson, Elaine Gan, and Nils Bubandt. eds. *Arts of Living on a Damaged Planet: Ghosts of the Anthropocene*. University of Minnesota Press, 2017.

Tucker, Herbert F. "In the Event of a Second Reform." *BRANCH: Britain, Representation and Nineteenth-Century History*. June 2012. http://www.branchcollective.org/?ps_articles=herbert-f-tucker-on-event.

Turpin, Etienne. "The Same River Twice: Nature, Media, and the Philosophy of the Anthropocene." Public lecture, Carleton University, Ottawa, April 12, 2017.

Tyrell, Alex. "Samuel Smiles and the Woman Question in Early Victorian Britain." *Journal of British Studies* 39, no. 2 (2000): 185–216.

Wald, Priscilla. "Science and Technology." In *Companion to Critical and Cultural Theory*, edited by Imre Szeman, Sarah Blacker, and Justin Sully, 403–18. Wiley-Blackwell, 2017.

Watt-Cloutier, Sheila. *The Right to be Cold: One Woman's Story of Protecting Her Culture, the Arctic and the Whole Planet*. Penguin, 2015.
Weber, Samuel. *Benjamin's -abilities*. Harvard University Press, 2008.
Weisman, Alan. *The World Without Us*. Picador, 2007.
Welsh, Alexander. *George Eliot and Blackmail*. Harvard University Press, 1985.
West-Pavlov, Russell. *Temporalities*. Routledge, 2012.
Weston, Kath. *Animate Planet: Making Visceral Sense of Living in a High-Tech Ecologically Damaged World*. Duke University Press, 2017.
White, Hayden. "Historical Emplotment and the Problem of Truth." In *Probing the Limits of Representation: Nazism and the "Final Solution,"* edited by Saul Friedlander, 37–53. Harvard University Press, 1992.
Whitman, Walt. "Song of Myself." *Leaves of Grass*. 1891–92. https://whitman archive.org/published/LG/1891/poems/27
Whyte, Kyle. "Indigenous Science (Fiction) for the Anthropocene: Ancestral Dystopias and Fantasies of Climate Change Crises." *Environment and Planning E: Nature and Space* 1, no. 1–2 (2018): 224–42.
Wiesenfarth, Joseph. *"Middlemarch*: The Language of Art." *PMLA* vol. 97, no. 3 (1982)" 363–77.
Williams, Daniel. "The Clouds and the Poor: Ruskin, Mayhew, and Ecology." *Nineteenth-Century Contexts* 38, no. 5 (2016): 319–31.
Williams, Karel. *From Pauperism to Poverty*. Routledge and Kegan Paul, 1981.
Williams, Raymond. "Ideas of Nature." In *Problems in Materialism and Culture*, 67–85. Verso, 1980.
———. *Marxism and Literature*. Oxford University Press, 1977.
Willis, Martin. "Scientific Self-Fashioning after *Frankenstein*: The Afterlives of Shelley's Novel in Victorian Sciences and Medicine." *Nineteenth-Century Contexts* 41, no. 3 (2019): 321–35.
Wolfson, Susan J. "Table of Dates." In *Frankenstein; or, the Modern Prometheus*, by Mary Shelley, xxiii–xxxii. Edited by Susan J. Wolfson, Longman, 2007.
Wood, Gillen D'Arcy. *Tambora: The Eruption that Changed the World*. Princeton University Press, 2015.
Wood, James. *Serious Noticing: Selected Essays, 1997–2019*. Farrar, Straus, and Giroux, 2020.
Woolf, Virginia. *A Room of One's Own*. Harcourt Brace Jovanovich, 1957.
———. *The Waves*. Granada, 1982.
Wright, Julia M. "Pathologizing Procrastination: Or, the Romanticization of Work." *English Studies in Canada* 34, no. 2–3 (2008): 16–20.
Wynes, S., and Nicholas, K. "The Climate Mitigation Gap: Education and Government Recommendations Miss the Most Effective Individual Actions." *Environmental Research Letters* 1, no. 7 (2017): 1–9.
Wynter, Sylvia, and Katherine McKittrick. "Unparalleled Catastrophe for Our Species? Or, to Give Humanness a Different Future: Conversations." In *Sylvia*

Wynter: On Being Human as Praxis, edited by Katherine McKittrick, 9–89. Duke University Press, 2015.

Yeazell, Ruth Bernard. *Art of the Everyday: Dutch Painting and the Realist Novel.* Princeton University Press, 2007.

Zemka, Sue. *Time and the Moment in Victorian Literature and Society.* Cambridge University Press, 2011.

Žižek, Slavoj. *Living in the End Times.* Verso, 2010.

Index

academic writing on climate change. *See* forms; interruption; thinking
Adair, Gilbert, 137
Adam Bede (Eliot): on age of transition, 27, 204n9; "In Which the Story Pauses a Little" (Eliot), 30–31, 204n7; and mediation, 47; new form, 46–47; pause in, 134; period of publication, 162, 206n19; posttime, 27, 31–32; as realist novel, 30, 46; and slow time, 204n9
Adorno, Gretel, 91
afterlives, 34, 86(b3), 91(b3), 93–94(b2), 205n13
"After the Arctic Sublime" (Morgan), 205n12
after-time, 32–33, 40, 45, 175, 205n13
against the grain: bands of familiar stories, 24, 93(b3); Benjamin, 28, 50; caesura, 15; documentaries, 34; *Frankenstein*, 173; Mayhew, 29, 34, 36–37, 42, 50; *Middlemarch*, 128; Mosse, 29, 34, 42, 46, 50

Agamben, Giorgio, 29, 43–44, 75, 202n29, 207n33, 210n27, 211n27
agricultural time, 27, 45, 204n9
Allegra and Claire, friends of Percy Shelley, 89–90(b2)
Amazon, the rain forest, 5, 13, 18
Amazon, the store, 18
amber alerts, 209n11
Anderson, Christopher Gangadin, 206n20
angels: Benjamin's and Mosse's, 53; Benjamin's angels of history, 10–11, 23, 72–74, 179, 201n16; like ghosts, 48
angelus novus, 25, 165, 166
Angelus Novus (Klee), 9–10, 17, *20*, 72–73, 203n33
Animate Planet (Weston), 7
Anthropocene, 12, 49, 168, 171–72, 187, 197n2
Anthropocene Reading (Menely and Taylor), 81–82
Anthropocene: The Human Epoch (Burtynsky), 30

242　Index

anti-BIPOC racism, 197n15
The Arcades Project (Benjamin), 10, 11, 16, 83, 123–25(b1), 200n4, 208n41, 211n23
Archangel *(Frankenstein)*, 165, 166
Arendt, Hanna, 44
"Artist's Statement" *(Incoming,* Mosse), 47
Arts of Living on a Damaged Planet (Tsing), 7
"An Attempt at a 'Compositionist Manifesto'" (Latour), 10
Austin, J. L., 56–58, 63, 73–74

bad weather, blank, 166–72
bands: about, 119–25(b5); as braided, 109–10(b5); as striated time, 88(b5)
Baucom, Ian, *History 4° Celsius,* 10
Beck, Ulrich, *The Metamorphosis of the World,* 10
Beer, Gilliam, 222n7
Benjamin, Andrew, 11–13, 16, 25
Benjamin, Walter: and the afterlife, 94(b2); angels of history, 10–11, 23, 72–74, 78, 179, 201n16; clinging to normal, 67; commentary *versus* translatability, 16, 21; and crystal, 104(b5); and dialectical images, 35, 48; and emergencies, 21, 68, 74–75, 210n27; and form, 9, 11, 16–17, 20–21, 111–12(b5), 179; against the grain, 28, 50; and history, xvi, 19, 21, 24, 28, 69, 100(b3), 205n16; and interruption, 11–12, 25, 55–56, 70, 169; juxtaposing quotations, 145; and Klee's painting, 17, 22–23, 25; on Marx, 207n40; and Mosse, angels, 53; and performative work, 69, 74; and pure language, 15–16; and religion, 203n33; rock record as metaphor, 83; and Scholem's poem, 74; sitting in internment camp, 98(b3); speeding train, 55, 65, 69; and the Tay Bridge disaster, xii, xiv, 72, 183; on temporality, 71; on thinking, 16, 48; *Arcades Project,* 10, 11, 16, 83, 123–25(b1), 200n4, 208n41, 211n23; "Fire Alarm," 71; *One-Way Street,* 70–71, 210n21; "On the Concept of History," 18–19 (see also "Theses on the Philosophy of History"); Paralipomena, 18, 21
BIPOC communities, 197n15, 209n5
Bjornerud, Marcia, 82–84, 94–96(b5), 101–2(b5), 171, 220n22
Black people, 202n24, 209n5
Blair, Sara, 30
Blanchot, Maurice, 102–3(b3), 105–9(b3), 107–9(b6), 113(b3), 120–22(b3)
blank, 166–72
Bleak House (Dickens), 125(b3)
blue books, 36–37, 206n21
Bolsonaro, Jair, 5
Bourdieu, Pierre, 201n12
Braiding Sweetgrass (Kimmerer), 92–96(b1), 221n2
Brand, Dionne, 45, 197n14, 222n5
Brazil, 5
Brecht, Bertolt, 9
bridge. *See* Tay Bridge disaster
"The Bridge by the Tay" (Fontane), xiii
Bringhurst, Robert, *Learning to Die* (with Zwicky), 82
British Museum, London, UK, 95(b4)
Brooks, Peter, 159
brushing against the grain. *See* against the grain
Buckel, David, 26
Buckland, Adelene, 83, 98–99(b5)
Buck-Morss, Susan, 12, 25

Burtynsky, Edward, *Anthropocene: The Human Epoch*, 30
Butler, Judith, 146, 160, 161

caesura, 16, 48
Callaway, Elizabeth, 202n29
camera, Mosse's, 45, 52, 207n34
capitalism, 6, 11, 61, 74, 202n23
carbon economy, 4, 40–41
carbon emissions, 74, 129, 222n8
Carlyle, Thomas, 36
Carson, Rachel, 96–103(b6), 100–102(b1), 104–12(b1)
Casaubon *(Middlemarch)*: about, 127; advice to cousin Will, 130; description of himself, 129; on Dorothea pointing out that he should start writing, 136–37; and German language, 139, 215n22; has a fit in the library, 142; incapable of transitioning from note-taking, 131; "Key to All Mythologies" (Casaubon, *Middlemarch*), 128–29, 131–34, 138–40, 143; motive for writing, 215n24; proposal and marriage to Dorothea, 132, 138; sensitive to pity, 140–41; and sexual performance, 139, 214n20; shares work with Dorothea, 143, 144, 215n29; on things left undone, 135; work in terms of weight, 134, 214n15. *See also* Dorothea *(Middlemarch)*
Castranovo, Russ, 205n16
Central Park Five, 8, 33–34
Chakrabarty, Dipesh, 10, 114–15(b5), 204n3
Choi, Tina Young, 204n9
chrononormativity, 205n11
Chthulucene, 7, 49
Clark, Timothy, 197n3, 207n37
climate action: Are we doing now what we have spoken of? Will you not now do the work?, 136, 137; Climate Action Lesson (CAL) #1: *Be wary of trying to solve the climate crisis.*, 128–29; CAL #2: *Divide your activism into doable chunks.*, 129–31; CAL #3: *Know when to stop gathering information and start action.*, 131–32; CAL #4: *Be wary of self-constructed as well as collective-constructed delays.*, 132–33; CAL #5: *Do not be afraid to act now.*, 133–36; CAL #6: *Do not work in isolation; talk to others, share ideas.*, 136–38; CAL #7: *Be wary of the push and pull between doubt and grandeur in one's actions.*, 138–40; CAL #8: *Be wary of requiring ideal conditions in which to work.*, 140–42; CAL #9: *Do act on climate change as if you might die tomorrow.*, 142–44; CAL #10: *Make it social; work with a buddy, build a community.*, 144; CAL #11: Know when to take a break., 144
climate change: and academics, 6–8, 157, 161–62; beginning of thinking on, 204n3; compared to the climate change idea, x, 74, 186, 197n1; deniers of, 64, 136; as exponential and nonlinear, 41; and *Frankenstein*, 157, 171–72, 179, 187; *Incoming* (Mosse), 42; literature on, 220n17; and post-time, 159–60; responses to, 178; term usage, 201n6; thinking as action, 4–6; warnings of, 58–60, 65, 67, 210n19; and Woolf, 118–20(b4)
coal, 40
cognitive dissonance, 56, 210n17
Cognito, Ann, ix–xi, xv–xvii, 189, 210n25
Cohen, Jeffrey Jerome, 83, 208n44

"Colonizer, Interrupted" (Richardson), 197n7
Comay, Rebecca, 71, 210n23
common lands, UK, 36, 206n30
constative statements, 57, 64, 73, 209n8. *See also* performative statements
COVID-19, xvi–xvii
Crary, Jonathan, 204n1
Creature *(Frankenstein)*: and bad weather, 166, 169–70, 220n21; as blank/unnamed, 167; at the end, 156, 174–77, 179–80, 221n27; and Frankenstein, 158, 170, 184; imprint of, 171–72; naming of, 218n1; a second one, 183; sledge appears and disappears, 164, 220n21; and Walton, 177, 178
"Crossing Brooklyn Ferry" (Whitman), 22
Crystal Palace, London, 12
cuckoo's eggs, 189, 222n10

Darwin, Charles, 83, 222n7
Davis, Lydia, 105(b3)
Death Sentence (Blanchot), 105–9(b3)
"A Defence of Poetry" (Percy Shelley), 91–96 (b6)
Defoe, Daniel, *Robinson Crusoe*, 109–12(b6)
De Laceys *(Frankenstein)*, 221n28
de la Durantaye, Leland, 56, 210n20
democracy, 5, 197n15
Derrida, Jacques: connection to Shelley, Woolf, Eliot, 117–18(b5); and formal experimentation, 201n10; internal drifting of frames, 160; on "The Purloined Letter" (Poe), 159, 166–67, 219n8; "Derrida Didn't Come" (Adair), 137; "Living On/Border Lines," 24, 86–125(b3), 107–9(b6)
dialectical images, 11, 16, 35, 48–49, 74, 173, 200n4, 207n40

diary genre, 163, 220n17
Dickens, Charles, *Bleak House*, 125(b3)
documentaries, 29–30, 34–35, 44, 207n37, 208n42
Dorothea *(Middlemarch)*: about, 127, 128; arrival at Lowick and interruption, 138, 214n17; on Casaubon's work, 128–29, 139; as procrastinator's enemy/does not want to impede Casaubon's work, 132–33, 135–37, 138–39; sexual experience with Casaubon, 214n20; weeping from dispute with Casaubon, 133–35, 213n14; works with Casaubon, 141, 143–44, 215n29. *See also* Casaubon *(Middlemarch)*
Dowden, Edward, 86–90(b2), 99(b2)
dream, 102–3(b2)
Duncan, Ian, 204n9
Dutch painters, 204n6
DuVernay, Ava, 8, 33
Dyer, Geoff, 214n19

ecocide, 29, 82, 107(b4), 201n6
Ehrmann, Jacques, 97–98(b3)
Eliot, George: on Casaubon, 215n24; connection to Shelley, Woolf, Derrida, 117–18(b5); on Dorothea weeping, 135; and Dutch painters, 204n6; and leisure time, 33; and parables, 213n7; and post-time, 23, 51, 160; on procrastination, 24, 126, 203n1; and representational models, 208n41; used for pithy remarks, 127, 213n6; *Felix Holt*, 214n16; "In Which the Story Pauses A Little," 30–31, 204n7; *Wise, Witty, and Tender Sayings in Prose and Verse Selected from the Works of George Eliot* (Main), 127. *See also Adam Bede* (Eliot); *Middlemarch* (Eliot)
ellipses: attention to, 190, 212; Brand's use of, 222; Derrida's use of,

115(b3); as gaping hole, xiii, 78; Lowy's use of, 76; Woolf's use of, 24, 87–92(b4), 88–125(b4), 96–101(b4), 111–12(b4), 115(b4), 121–25(b4)
employment: *versus* environment, 5; *versus* technology, 37
Enclosure Acts, UK, 36, 50, 206n30
Englander, Nathan, 26
Enlightenment, 32, 220n16
Enright, Anne, 178
Entin, Joseph B., 30
environment *versus* employment, 5
Epidemic Illusions (Richardson), 197n7
epigraphs, 86
epoch of diachronicity (Malm), 210n16
equality/inequality, xv, 186, 199n15
Extinction Rebellion, 185, 215n30

Facing Gaia (Latour), 62, 209n4
failure, 187–88
The Fall (Mosse), 49
Felix Holt (Eliot), 214n16
Ferris, David, 11–12, 25, 201n17, 203n35
feuilleton, 210n22
Fire Alarm (Löwy), 10, 76
"Fire Alarm" (Benjamin), 71
Flear, Noam, 84
Foer, Jonathan Safran, 201n10
Fontane, Theodore, xiii
Ford, T. H., 162, 219n12
forms: academic writing on climate change, 6–8; *Adam Bede* (Eliot) as new form of storytelling, 46; and Benjamin, 9, 11, 16–17, 20–21, 111–12(b5), 179; books in different forms, 6–7, 13; changing form to change time, 23, 25; Derrida experimentation with, 86(b3); documentaries as, 30; experimental, 179; familiar, 84; and interruptions, 76, 85, 111–12(b5); invisibility of cultural or academic, 81; letters and time, 164; *London Labour* as example of different forms of print, 39; meaning through form (Benjamin), 11, 16, 17, 20–21; narrative form of *Middlemarch* (Eliot), 214n21; new for expanding range of representation, 47; new visual and literary, 83; questions as, 146; of temporal experiences, 205n11; and thinking, 4–5, 116(b5); Tsing on, 15
Fossil Capital (Malm), 10, 65, 67
fossil fuel emissions, 54
Foucault, Michel, 117–25(b6), 162
framing strategies, 157, 158–62, 166–67, 176–80
Frankenstein. *See* Creature *(Frankenstein); Frankenstein* (Mary Shelley); Frankenstein, the character; Margaret *(Frankenstein);* Shelley, Mary; Walton *(Frankenstein)*
Frankenstein (Mary Shelley): about, 162–65, 219n1, 219n13, 221n1; afterlives of, 157, 219n11; Archangel, 165, 166; and bad weather, 169–70; break in the story, 172; and climate change, 157, 171–72, 179, 187; De Laceys, 221n28; foreshadowing, 175; as framed narrative, 24, 157–62, 176–80; Freeman on, 219n11; and ghost stories, 161; against the grain, 173; as inspiration, 162; teaching, 158; Mount Tambora, Indonesia, 171, 220n22; orientation to time, 24, 157; Percy Shelley Preface, 156; quotes from, 156, 158, 162, 176; readers of, 167; St Petersburgh, 163, 165, 219n13, 219n14; on technologies, 157–58, 171–72; junstableness of, 178–79
Frankenstein, the character: and the Creature, 171–72, 174, 177, 221n28; final words, 172–73; as ghost, 172–76, 177, 221n28; responses

Frankenstein (continued)
 to climate change/Creature, 168;
 and a second creature, 183–84,
 221n1; sickens, 166, 220n20; to
 Walton, 156, 158, 164–66, 220n20.
 See also Creature *(Frankenstein);*
 Frankenstein (Mary Shelley);
 Margaret *(Frankenstein);* Walton
 (Frankenstein)
Freeman, Elizabeth, 205n11, 219n11
Freeman, Michael, 210n15
Fridays for Future, 215n30
Fuel: A Speculative Dictionary
 (Pinkus), 7
Future Rising, 215n30
Fyfe, Paul, 210n24

gaping holes: about, ix–xvii; always
 there, xiv; blank/unnameable, 167,
 169, 170–71; of the bridge over the
 River Tay, 179; and climate disruption, xiv, 74; documentaries and,
 30; and *Frankenstein,* 158, 170, 187;
 how to act when looking at, 188;
 and the humanities, 5; interruptions as, 13–15; and reconstellation
 of time, xvi; staying with, 145, 178,
 187; Tay Bridge disaster, 183; and
 the Twin Towers, 9/11, 7
Gardiner, Stephen, *A Perfect Moral*
 Storm, 66
Geological Survey of Ontario,
 90–92(b5)
The Geological Unconscious (Groves),
 201n16
geology, 82, 83, 91–93(b3)
Ghosh, Amitav, 201n20
ghosts: and angels, 48; in *Bleak House,*
 Hamlet, 125(b3); and *Frankenstein,*
 172–76, 177, 221n28; Mary Shelley
 writings, 94(b3), 161, 185–86; as
 mediation, 95(b3); of other texts,
 95(b3); stories, *Frankenstein* (Mary
 Shelley), 161
Girl, Interrupted (Kayson), 197n7
"Girl Interrupted at Her Music" (Vermeer), 197n7
Girton College, 92(b4)
Gisborne, John, 96(b2)
Global North: cognitive dissonance
 for, 56; default responses, 14, 168–
 69; political configuration of, 69,
 77; reading in bands, 88–89(b5); reports instead of action, 132; separa-
 tion of ideas, 115–16(b5); temporal
 thought, x–xi, xiv, 203n35; and
 typical action, 187, 222n8; as "we,"
 199n8, 209n7
global warming, 201n6
Gore, Al, *An Inconvenient Truth,* 29–30
Gould, Stephen Jay, 197n10
grain, against the. *See* against the grain
Grain Vapour Ray (Klingan), 7
Gramineæ, 120–23(b1)
grasses, 86, 90
Great Exhibition, London, 12–13
greenhouse gas emissions, 3
Green New Deal, 215n30
Groves, Jason, *The Geological Uncon-*
 scious, 201n16
Guillory, John, 204n10

Haight, Gordon S., 214n20, 215n22
hallucinations, 94(b3)
Hamlet (Shakespeare), 89(b3),
 118–19(b3), 125(b3)
Haraway, Donna, 7, 49
Harrison, Robert Pogue, 82
heat images, 44
Heat Maps (Mosse), 24, 28, 42
Heidegger, Martin, 117(b3), 145, 146,
 169, 206n30
Henry VI (Shakespeare), 87(b6)
Herzon, Werner, 48

History 4° Celsius (Baucom), 10
hockey stick graph, 63, 209n8
Holocaust, 29, 44
Homer, 45, 46
Homo Sacer (Agamben), 43–44
house on fire, 62, 63, 68, 209n11. *See also* Thunberg, Greta
How Forests Think (Kohn), 13
How to Do Things with Words (Austin), 57, 73
Hunt, Leigh, 166, 167
Hunt, Marianne, 105–6(b2)
Hutton, James, 83
Hyde, Lewis, 84

Illustrated London News, 210n24
Incoming (Mosse), 24, 28, 42–44, 47–53, 188
An Inconvenient Truth (Gore), 29–30
indecision, 88(b3), 89(b3), 94–96(b3), 99(b3)
"Indigenizing the Anthropocene" (Todd), 209n7
Indigenous Peoples, 14, 189, 197–98n4, 203n34, 203n35, 209n5, 221n2
industrial modernity: and agricultural time, 27, 204n9; Benjamin's response to, 72, 83; and common lands, 36; documented by *London Labour,* 28, 47; and fossil fuel economy, 65; invoked by gaping hole in train line, xii; and linear time, 23, 203n35; problems covered up by images, 210n24; production and time, 32; and rise of the Anthropocene, 12
Industrial Revolution, xi, 68, 126, 198n4, 220n16
Intergovernmental Panel on Climate Change Report (2021), 3, 132, 143
interruption/interruptions: academic writing as, 8–11, 183–89;

achievement of, 19; arriving, waiting, 162–66; Benjamin on, 11–12, 25, 55–56, 70, 169; "Colonizer, Interrupted" (Richardson), 197n7; and distraction, 11; enabling or disabling, 120(b4); facilitated by writing and thinking, 25; Ferris on, 11–12; and form, 76, 85, 111–12(b5); as gaping hole, 13–15; "In Which the Story Pauses A Little" (Eliot), 30–31, 204n7; as performed, 55; returning, or the ghost of Frankenstein, 172–76; in *A Room of One's Own,* 93–94, 93–95(b4); in sustained thinking, 111(b5); *versus* warnings, 55; and women, 97–99. *See also Adam Bede* (Eliot); ellipses; Frankenstein; Mayhew, Henry; *Middlemarch* (Eliot); Mosse, Richard; procrastination; questions; Shelley, Percy; sitting/staying with; Tay Bridge disaster; Thunberg, Greta; time; Woolf, Virginia
"In the State of Emergency" (Malm), 67
In the Wake (Sharpe), 205n13
"In Which the Story Pauses a Little" (Eliot), 27, 30–31, 204n7

James, Henry, 204n1
Jennings, Michael, 71
Johnson, Barbara, 159–60, 219n7, 219n8
Joyce, James, *Ulysses,* 46

Kayson, Susan, *Girl, Interrupted,* 197n7
Keat's poetry, book of, 86(b2)
"Key to All Mythologies" (Casaubon, *Middlemarch*), 128–29, 131–34, 138–40, 143
Kimmerer, Robin, 92–96(b1), 221n2
Kingwell, Mark, 137, 213n8, 213n12

Klee, Paul, 9–10, 17, 18, *20*, 72–73, 203n33
Klein, Naomi, 55, 60–63, 68, 75, 102–4(b1), 209n7
Klingan, Katrin, *Grain Vapour Ray*, 7
Kluge, Alexander, 22
Kohn, Eduardo, *How Forests Think*, 13
Krenak, Ailton, xv

Lacan, Jacques, 159
Latour, Bruno: "An Attempt at a 'Compositionist Manifesto,'" 10; appeal to the burning house, 55, 78; on bringing people/disciplines together, 200n5, 202n20, 209n7; climate warnings not working, 60, 62–63, 75, 209n6; could have acted . . . , 210n17; *Facing Gaia*, 62, 65, 209n4; future defines the present, 207n30; from panic to action, 209nn9–10; and performative statements, 64, 73, 209n8; under tension, 63, 64, 69, 210n11; *We Have Never Been Modern*, 209n4
Learning to Die (Bringhurst and Zwicky), 82
Leaves of Grass (Whitman), 97–100(b1)
leisure time, 27–28, 31, 204n8
Lemenager, Stephanie, 90–91(b6), 203n32
Lerner, Ben, 22
Lesjak, Carolyn, 206n30
Levine, Caroline, 31, 138
Life of Percy Bysshe Shelley (Dowden), 88–89(b2)
Linnæus, 86(b5)
literary studies, 82
"Living On/Border Lines" (Derrida), 24, 86–125(b3), 107–9(b6)
logos, 82
London Labour and the London Poor (Mayhew): about, 12–13, 23–24, 28; and blue books, 36, 37; compared to Benjamin's *Arcades*, 208n41; as documentary/exposé, 28, 35; editorship of, 38, 205n18; effect of palimpsest, 38; etymologies in, 39–40; Mayhew's commentaries, 40, 206n29
London Vagabond (Anderson), 206n20
Long Soldier, Layli, 96–97(b1)
Lovitt, William, 146
Löwy, Michael, 10, 70, 76, 207n40
Luther, Martin, 22, 203n33
Lyell, Charles, 83, 86(b5), 87–88(b6), 112–19(b1), 197n10

Macfarlane, Robert, 53, 83, 100–101(b5)
Madness of the Day (Blanchot), 102–3(b3)
Main, Alexander, 127
Malm, Andreas, 10–11, 55, 65, 67, 69, 75, 210n16
Mann, Michael, 63
Margaret *(Frankenstein)*: evil forebodings, 163, 165; as recipient of Walton's letters, 158–59, 161, 166, 177
Marx, Karl, 29, 70, 102(b5)
Mayhew, Henry: about, 206n26; new form to capture the impact of industrial modernity, 13, 46–47; and post-time, 36; sees work as encyclopedia of London, 206n21; temporal polyphony, 34; and time and mediation, 51; and work as documentary reportage, 34, 35, 37; work compared to Mosse, 28–30, 41–42, 49–51. *See also London Labour and the London Poor*
McCole, John, 71
McGonagall, William, xii–xiii
McKittrick, Katherine, 7
"Meaning to Get To" (Kingwell), 213n8

mediation: and *Adam Bede*, 47; and bands, 110(b5); in both Mosse and Mayhew, 50; concept found in Derrida, 121(b3); and documentaries, 30; ghosts as, 95(b3); *London Labour* as case study, 39, 40; and Mosse and Mayhew, 51; and post-time, 178; of punctuation, 87(b4); and rocks, 87(b5); stories framed, 158; and time, 33; underscored by Mosse's slow technology, 208n43
men, writing about women, 96(b4)
Menely, Tobias, 81–82, 83, 197n2, 201n16
Menke, Richard, 32
The Metamorphosis of the World (Beck), 10
Michelet, Jules, 42
Middlemarch. *See* Casaubon *(Middlemarch);* Dorothea *(Middlemarch);* Eliot, George; *Middlemarch* (Eliot); procrastination; procrastination lessons from *Middlemarch*
Middlemarch (Eliot): about, 126–27, 144; narrative form of, 214n21; and procrastination, 24, 126
migration, 42, 49–50
Miller, Andrew, 185, 221n4
Miller, D. A., 214n21
modernism, 203n35
Moore, Jason, 204n4
Morgan, Benjamin, "After the Arctic Sublime," 205n12
Morning Chronicle, 35, 36, 38, 40
Morton, Timothy, 29
Mosse, Richard: camera, 45–46, 52, 207n34; and documentaries, 34, 208n42; illuminations of temporalities, 45; on the medium, 47–48; slowing down, 50, 207n36, 208n43; temporal polyphony, 34; and time and mediation, 51; work compared to Mayhew, 28–30, 34, 41–42, 49–51; *Heat Maps*, 24, 28, 42; *Incoming*, 24, 28, 42–44, 47–53, 188
Mount Tambora, Indonesia, 171, 220n22
Mullen, Mary, 34
Mushroom at the End of the World (Tsing), 6–7

natural laws, 208n40
neoliberalism, 11, 61, 64, 68, 69, 77
9/11, 7, 13, 64, 209n11
Nixon, Rob, x–xi, 29, 210n18
novels, realist, 27, 30
Nudelman, Franny, 30, 50, 207n36

Occupy Movement, xv
Odell, Jenny, 49
Odyssey (Homer), 45, 46
Olmsted, Frederick Law, 7
1 percent, xv
One-Way Street (Benjamin), 70–71, 210n21
On Fire (Klein), 61
"On the Concept of History" (Benjamin), 18–19
The Order of Things (Foucault), 117–25(b6)
The Origin of the Species (Darwin), 206n19
Oxford English Dictionary (OED), 120–23(b1), 213n8
Oxford Junior Dictionary, 100–101(b5)

palimpsests, and quality of: bands, 92(b3); and Bjornerud, 82; *London Labour as*, 38; and post-time, 41, 118(b5), 160–61, 178, 188; and time, 32–33, 50, 83, 204n10
pandemic, xvi–xvii
panic, 65, 68, 77
Paralipomena (Benjamin), 18, 21

Parerga/parergon, 139, 141, 215n25
Parkes, Bessie Rayner, 36
pastime/past-time, 31, 41, 204n8
A Perfect Moral Storm (Gardiner), 66
performative statements: Benjamin's work, 74–75; and constative, 57, 64, 73, 209n8; contradictions, 55, 69, 73, 128; description as, 61; speech acts, 67; success of, 208n2
periodicity, 27–28, 204n2
Pinkus, Karen, *Fuel: A Speculative Dictionary,* 7
Poe, Edgar Allen, 159–60, 219n9
post, term usage, 32, 205n14
postmodernism, 203n35
post-time: in *Adam Bede* (Eliot), 31–32; as after-time, 45; age of transition, 27; and Derrida's reading of *Death Sentence,* 107(b3); as Eliot's term, 23, 28, 36; and Mosse and Mayhew, 51; offers possibilities, 34; as palimpsestic, polytemporal, 41, 188; and progress, 34–35; temporal polyphony, 34, 51, 160, 178; term usage, 28, 33, 204n2
potato famine, 50
Price, Leah, 127, 213n6
Principles of Geology (Lyell), 86(b5), 87–88(b6), 112–19(b1)
procrastination: children as, 138–39, 214n19; and climate crisis, 127; definitions, 126, 213n8; and feedback, 137, 214n18; and *Middlemarch,* 24, 126, 127; procrastinator's dilemma, 139, 140, 143, 214n21; requires a procrastinator, 138. *See also* procrastination lessons from *Middlemarch*
procrastination lessons from *Middlemarch*: Procrastination Lesson (PL) #1: Be wary of aspiring to greatness., 128–29, 136, 137, 140–41; PL #2: Divide your project into doable chunks., 129–31, 136; PL #3: Know when to stop taking notes and start writing., 131–32, 135, 136; PL #4: Be wary of self-constructed delays and distractions., 132–33, 138; PL #5: Do not be afraid to act now., 133–36; PL #6: Do not keep your work to yourself; share it with others, even "ignorant onlookers.," 136–38; PL #7: Be wary of the procrastinator's dilemma., 138–40; PL #8: Be wary of requiring ideal conditions in which to write., 140–42; PL #9: Do write as if you might die tomorrow., 142–44; PL #10: Make it social; work with a buddy, build a community., 144; PL #11: Know when to take a break., 144
procrastinator's dilemma, 139, 140, 143, 214n21
The Progress of This Storm (Malm), 10, 66
Prospect Park, Brooklyn, 26
prussic acid, 105(b2)
punctuation, 86–87(b4). *See also* ellipses
"The Purloined Letter" (Poe), 159–60, 219n9
Pychyl, Timothy A., 214n18

"Queen Mab" (Percy Shelley), 89–90(b6)
questions: on agency, 149; on being late to respond, 149–50; on the end, 155; as form, 145–46; on habitat destruction, 146–47; on nature, 148–49; on other realities, 153–54; on stories of home, 151–53; unanswerable, 147–48; on wealth and impact, 154–55; on what to do, 150–51

rain forest, 5, 13, 18
Ranciere, Jacques, 205n16

Ranke, Leopold von, 47
realism, 29–30, 47, 207n37
refugees, 28, 45, 49
Remaking Reality (Blair, Entin, and Nudelman), 30
Richardson, Eugene, *Epidemic Illusions,* 197n7
River Tay. *See* Tay Bridge disaster
Robinson Crusoe (Defoe), 109–12(b6)
rocks/stones, 81, 86–125(b5)
A Room of One's Own (Woolf), 87–125(b4)
Rudwick, Martin, 83
Ruefle, Mary, 84

Sachs, Jonathan, 203n1, 204n9
Said, Edward, 197n1
Savoy, Lauret, 4, 86–87(b6), 98–100(b5)
Scholem, Gershom, 17
"Science and Technology" (Wald), 10
Scott, Walter, 171
Scranton, Roy, 14–15
The Sea Around Us (Carson), 96–103(b6), 100–102(b1), 104–12(b1)
Secord, James, 197n10
Sedgwick, Eve, 34
"Sediment" (Lemenager), 90–91(b6)
Self-Help (Smiles), 126
Shakespeare: *Hamlet,* 89(b3), 118–19(b3), 125(b3); *Henry VI,* 87(b6)
Sharpe, Christina, 197n14, 202n24, 205n13, 222n5
Shelley, Mary: about, 24; on the banks of the River Tay, 183; on blankness/unnameable, 167–69; and ghost stories, 185–86; letter to Leigh Hunt, 166; and Percy, 92(b2), 100–103(b2), 113–17(b2), 120–21(b2); writing after the revolutions, 220n16. *See also Frankenstein* (Mary Shelley)

Shelley, Percy: "A Defence of Poetry," 91–96(b6); building a boat, accident, and death, 24, 87, 98–99(b2), 109–11(b2); connection to Woolf, Eliot, Derrida, 117–18(b5); cremation, 92(b2); dream, 102–3, 124(b2); historical play of Charles I, 95(b2); and linear time, 96(b2); plans for dying, 104–6(b2); preface to *Frankenstein,* 156; premonition of death, 90–91(b2); "Queen Mab," 89–90(b6); on roads/rivers, 220n19; seeing Allegra in the water, 89(b2); sense of time, 106–7(b2); "The Triumph of Life," 87–88(b3), 93–94(b3), 93–95(b2), 99(b2), 108(b2), 124(b3)
Sinnema, Peter, 126, 210n24
Siskin, Clifford, 32
sitting/staying with, xiv, 15, 197n14, 202n24
Sloterjik, Peter, 14–15
slowness, x, xi, 29, 31, 204n9, 207n36
Smiles, Samuel, 126
Smithson, Robert, 83–84, 103–4(b6)
Smithsonian, 91(b5)
The Snows of Venice (Lerner and Kluge), 22
Solar Radiation Management, 171, 220n22
Solvo, Pablo, 60
Sörlin, Sverker, 104–7(b6)
speeding train, 65, 66, 67, 69, 76
Spivak, Gayatri, 159, 177
Staying With the Trouble (Haraway), 7
Stephenson, George, 72
stones/rocks, 81, 86–125(b5)
St Petersburgh *(Frankenstein),* 163, 165, 219n13, 219n14
suicide, 82, 107(b4), 125(b4)
superposition/superimposition, 92(b3)
Surowiecki, James, 213n8
surrealist experiments, 203n35

surveillance systems, 64
sustainability discourse, 14, 202n23

"The Task of the Translator" (Benjamin), 21
Tay Bridge disaster: about, xii–xiii, xii–xiv, xiv, 72, 179, 183; "The Bridge by the Tay" (Fontane), xiii; "The Tay Bridge Disaster" (McGonagall), xii–xiii
Taylor, Astra, 5
Taylor, Isaac, 96–98(b5)
Taylor, Jesse Oak, 41–42, 81–83
technologies: beginning to be felt, 12; Benjamin on, 72; change experience of time, 28, 32; *Frankenstein* on, 157–58, 171–72; Mayhew and Mosse, 37, 42–43, 45–47, 49–51; science and technology studies, 10, 200n5. *See also* camera, Mosse's
temporal polyphony, 34, 51, 160, 178
10:04 (Lerner), 22
tension, under, 63, 64, 69, 210n11
thermal radiation, 46
"Theses on the Philosophy of History" (Benjamin): about, 9–10, 73, 203n33; de la Durantaye on, 210n20; and interruptions, 18; on Klee's painting, 17, 21, 23; narratives of progress, 13; Paralipomena (Benjamin), 18–21; quote from, 86–92(b1). *See also* against the grain; Benjamin, Walter
The Thief of Time: Philosophical Essays on Procrastination (Main), 213n8
thinking: as action, 4–6; Benjamin on, 48; not reserved for humans, 200n6; other ways of, 221n2
This Changes Everything: Capitalism vs. The Climate (Klein), 60, 102–4(b1)
Thompson, Lonnie G., 61
Thunberg, Greta: beginning of movement, 216n30; call for a new politics, 61; inspired other climate protests, 185; "I want you to act," 56; "I want you to panic," 68; "Our house is on fire," 24, 54–56, 59–60, 78, 124(b2), 208n2; presentations, 76–78
time: after-time, 32–33, 40, 45, 175, 205n13; Bjornerud on, 82; changes in, 25, 28, 43; character and author in different times, 163, 220n16; deep, 82, 90(b5), 162; Derrida's reading of, 89(b3); disturbance to in mid-nineteenth century, 35–36, 206n19; and Indigenous Peoples, 189, 197–98n4, 203n34, 203n35; inhabit differently, 11; letters and form, 164; materiality of, 32; mediation of, 32; Mosse on, 46, 51; organization by narrative, 83; organization of (schedules, calendars, time zones, wristwatches), 44, 82, 205n11; as palimpsestic, 33; pastime/past-time, 31, 41, 204n8; remaking of, 158; role of, 8; slowness, x, xi, 29, 31, 204n9, 207n36; studies of, 197n4, 203n35. *See also* post-time
Timefulness (Bjornerud), 84, 96(b5)
Todd, Zoe, 209n7
Trace (Savoy), 4
trains, 65–66, 72, 210nn14–15
translatability, 16–17, 21–22, 101(b3), 145
Trelawney, Edward, 104(b2), 110(b2)
"The Triumph of Life" (Percy Shelley), 87–88(b3), 93–94(b3), 93(b3), 124(b3)
Tsing, Anna, 6–7, 15, 34
Tucker, Herbert, 41
Twin Towers, NYC, 7

Ulysses (Joyce), 46
Underland (Macfarlane), 53
under tension, 63, 64, 69, 210n11

"Unparalleled Catastrophe for our Species?" (Wynter and McKittrick), 7
urban pollution, 40–41

Vaux, Calvert, 7
Vermeer's "Girl Interrupted at Her Music," 197n7
video installation, 43
visual art, 201n12, 203n35
Vivian, Charles, 87(b2), 109–11(b2)

Wald, Priscilla, 10
"Walk to Waken the Nation" (Cognito), 210n25
Wallins, Roger, 36
Walton *(Frankenstein):* about letters to his sister, 158–61; and the Creature, 170, 178, 179–80; decision to return, 174, 220n25; discoveries of, 163, 220n15; Frankenstein's story, 164–65; interruption, 173, 174, 177; letters as frame, 176; letters to Margaret, 171, 172, 173, 176; in St Petersburgh, 163, 219n13, 219n14; and time, 162–63; timing of letters, 163–64, 165; waits out the storm, 166
Warner, William, 32
warnings, 55, 56–60, 63, 75
The Waves (Woolf), 112–17(b6)
"we," 198n8, 221n5
weapon, camera as, 46, 207n34
We Are the Weather (Foer), 201n10
weather: blank, 166–72; and humans and non-humans, 178
Weber, Samuel, 15–17, 21, 87(b4), 93–94(b2), 145, 202n27
We Have Never Been Modern (Latour), 209n4
Weisman, Alan, 201n14
Welsh, Alexander, 134, 214n16, 215n24
"We Shall Look Into It Tomorrow" (Kingwell), 213n8

Weston, Kath, *Animate Planet,* 7
Wet'suwet'en territory, 14
When They See Us (DuVernay), 8
whiling *versus* wilding, 8, 13, 18, 31, 33
White, Hayden, 29, 47
Whitman, Walt, 22, 97–100(b1), 98–99(b5)
wildfires, 13, 208n2
wilding *versus* whiling, 8, 13, 18, 31, 33
Williams, Edward, 87(b2), 98–99(b2), 109–11(b2)
Wise, Witty, and Tender Sayings in Prose and Verse Selected from the Works of George Eliot (Main), 127
Wollstonecraft, Mary, 177
women's rights, 114–17(b4)
Wood, Gellen D'Arcy, 171–72
Wood, Henry, 205n18
Wood, James, 221n4
Woolf, Virginia, 87–125(b4); and climate change, 118–20(b4); connection to Derrida, Shelley, and Eliot, 117–18(b5); use of ellipses/dots, 24, 87–92(b4), 96–101(b4), 111–12(b4), 115(b4), 121–25(b4); *A Room of One's Own* (Woolf), 87–125(b4); *The Waves* (Woolf), 112–17(b6)
Words and Places (Taylor), 96, 96–98(b5)
Wordsworth, William, 203n1
World Economic Forum, Davos, Switzerland, 54
Wynter, Sylvia, 7, 197n8, 199n8, 221n5

Yeazell, Ruth Bernard, 30–31
youth movements, 209n5, 210n28

Zemka, Sue, 204n9
Zerubavel, Evitar, 205n11
Zwicky, Jan, *Learning to Die* (with Bringhurst), 82

The authorized representative in the EU for product safety and compliance is:
Mare Nostrum Group
B.V Doelen 72
4831 GR Breda
The Netherlands

www.ingramcontent.com/pod-product-compliance
Lightning Source LLC
Chambersburg PA
CBHW022004220426
43663CB00007B/948